D L Colton
University of Strathclyde

Partial differential equations in the complex domain

Pitman Publishing
LONDON · SAN FRANCISCO · MELBOURNE

PITMAN PUBLISHING LIMITED
39 Parker Street, London WC2B 5PB

FEARON–PITMAN INC.
6 Davis Drive, Belmont, California 94002, USA

Associated Companies
Copp Clark Ltd, Toronto
Pitman Publishing Co. SA (Pty) Ltd, Johannesburg
Pitman Publishing New Zealand Ltd, Wellington
Pitman Publishing Pty Ltd, Melbourne
Sir Isaac Pitman Ltd, Nairobi

AMS Subject Classifications: 35C15, 35A20, 35R25, 35R30

First published 1976
Reprinted 1977

Reproduced and printed by photolithography
in Great Britain at Biddles of Guildford

ISBN 0 273 00101 9

Contents

Introduction

he subject matter of these lectures can conveniently be intro-uced by quoting a paragraph from Methods of Mathematical hysics, Vol. II by Courant and Hilbert: "The stipulation bout existence, uniqueness, and stability of solutions domi-ate classical mathematical physics. They are deeply inherent n the ideal of a unique, complete and stable determination of hysical events by appropriate conditions at the boundaries, t infinity, at time t = 0, or in the past. Laplace's vision f the possibility of calculating the whole future of the physi-al world from complete data of the present state is an extreme xpression of this attitude. However, this rational ideal of ausal-mathematical determination was gradually eroded by con-rontation with physical reality. Nonlinear phenomena, quantum heory, and the advent of powerful numerical methods have shown hat 'properly posed' problems are by far not the only ones hich appropriately reflect real phenomena. So far, unfortuna-ely, little mathematical progress has been made in the import-nt task of solving or even identifying and formulating such roblems which are not 'properly posed' but still are important nd motivated by realistic situations". In one sense these otes are an introduction to the use of function theoretic ethods in the investigation of one important class of physic-lly motivated "improperly posed" problems, that is improperly osed initial value problems. However, such a study extends ar beyond the immediate physical situation in which these roblems arise. The following example serves as an illustration: he uniqueness of a solution to the "improperly posed" elliptic auchy problem is equivalent to the Runge approximation roperty. In order to exploit such a property one is led to construct an integral operator which maps analytic functions onto solutions of the elliptic equation under investigation, thus giving a practical method of constructing both a complet family of solutions and analytic approximations to the ellipt Cauchy problem. However in order to construct this integral operator it is necessary (in the case of three independent va ables) to again examine an "improperly posed" problem, this time an exterior characteristic initial value problem for a hyperbolic equation. Having now constructed the desired inte gral operator it can in turn be used not only to construct a complete family of solutions but also to analytically continu solutions of the elliptic equation from a knowledge of the domain of regularity of their Cauchy or (complex) Goursat dat along prescribed analytic surfaces. Proceeding in this manne and considering selected "improperly posed" initial value pro lems it is possible to systematically develop an analytic the of partial differential equations based on the analytic theor of functions of a complex variable, and in a broad sense it i to this general theory that these notes are devoted.

We now briefly outline the content of the lectures. Chapt one consists of statements of basic results on the existence, uniqueness, and regularity of solutions to initial and bounda value problems in partial differential equations which will b needed in future chapters. In Chapter two we consider "improperly posed" initial value problems, give examples of their appearance in physics, and obtain results on the analyt continuation of solutions to elliptic and parabolic equations In particular, the result mentioned above on the equivalence of the uniqueness to the elliptic Cauchy problem and the Rung approximation property is proved. In Chapter three we constr integral operators which map analytic functions of one and several complex variables onto real valued solutions of ellip tic equations in two and three independent variables. Since this particular topic has been the subject matter of three different books ([1], [21], [39]), we concentrate on newer results, in particular integral operators for elliptic equati in three independent variables. Nevertheless for the sake of completeness we construct (in section 8) Bergman's integral operators for elliptic equations in two independent variables and derive (in section 11) Vekua's representation of solutions to this class of equation. Chapter four is concerned with the use of integral operators in the analytic continuation of solutions to elliptic equations. Lewy's reflection principle and Gilbert's envelope method are derived and as an example of their use are applied to the problem of analytically continuing solutions of the axially symmetric Laplace and Helmholtz equations. Included in this chapter is a discussion of radiation conditions and asymptotic expansions of solutions to the axially symmetric Helmholtz equation. In Chapter five we introduce a class of third order equations which have been the object of recent study in various areas of fluid dynamics and derive the basic analytic properties of solutions to such equations. It is shown that in terms of their analytic behaviour the solutions of the third order equations considered here occupy a position somewhere in between that of parabolic and elliptic equations. "Improperly posed" problems associated with these equations can be found in [5].

None of the material presented in the last four chapters of these notes has appeared in previous research monographs on the subject, except for section 4 (which can be found in Chapter 16 of [20]), section 8 (which is taken from [2] and can also be found in [1] and [21]) and section 12 (which is taken from [21]). This fact is reflected in the bibliography, where only those results which are directly referred to or used in these notes are referenced. For other work in this area the reader is directed to the bibliographies contained in the books by Bergman ([1]), Garabedian ([20]), Gilbert ([21]), and Vekua ([39]).

The following course of lectures was given in the spring semester of 1971 at Indiana University and again in the fall of that year at the University of Glasgow where the author was a visiting research fellow participating in the North British

Symposium on Partial Differential Equations and Their Applica tions. Gratitude is expressed to the Science Research Counci the National Science Foundation, and the Air Force Office of Scientific Research for their financial support, and to Professor Ian Sneddon and the University of Glasgow for their hospitality during the academic year 1971-1972 when the first draft of these notes was typed and circulated.

Note: Since the time these notes were written a considerable amount of new research has been completed in areas closely related to the subject matter of these lectures. For some of these new results the reader is referred to

R.P.Gilbert, Constructive Methods for Elliptic Partial Differential Equations, Springer Verlag Lecture Notes Serie Berlin, 1974.

I Preliminary results

Preliminaries

Let $\underset{\sim}{x} = (x_1,\ldots,x_n)$, $\underset{\sim}{\xi} = (\xi_1,\ldots,\xi_n)$ and let $u(\underset{\sim}{x})$ satisfy the equation

$$L[u] = \sum_{i=1}^{n} \frac{\partial^2 u}{\partial x_i^2} + \sum_{i=1}^{n} b_i(\underset{\sim}{x}) \frac{\partial u}{\partial x_i} + c(\underset{\sim}{x})u = 0$$

where $b_i(\underset{\sim}{x})$, $i = 1,\ldots, n$, and $c(\underset{\sim}{x})$ are analytic functions of their independent variables in a domain D.

1. Fundamental Solutions and Analyticity

A fundamental solution $S = S(\underset{\sim}{x};\underset{\sim}{\xi})$ is a solution of the equation

$$L[S] = \delta(|\underset{\sim}{x} - \underset{\sim}{\xi}|)$$

where δ denotes the Dirac delta function. S has the form

$$S = \frac{U(\underset{\sim}{x};\underset{\sim}{\xi})}{r^{n-2}} + V(\underset{\sim}{x};\underset{\sim}{\xi}) \log r + W(\underset{\sim}{x};\underset{\sim}{\xi});\ n > 2 \tag{1.1}$$

$$S = A(x_1,x_2;\xi_1,\xi_2) \log \frac{1}{r} + B(x_1,x_2,\xi_1,\xi_2);\ n = 2 \tag{1.2}$$

where $r = |\underset{\sim}{x} - \underset{\sim}{\xi}|$. For n odd $V \equiv 0$, $L[W] = 0$. In (1.2) set $z = x_1 + ix_2$, $z^* = x_1 - ix_2$, $\zeta = \xi_1 + i\xi_2$, $\zeta^* = \xi_1 - i\xi_2$. Then

$$\underset{\sim}{L}[A] = \frac{\partial^2 A}{\partial z \partial z^*} + \alpha\frac{\partial A}{\partial z} + \beta\frac{\partial A}{\partial z^*} + \gamma A = 0 \tag{1.3}$$

where $\alpha = \frac{1}{4}(b_1+ib_2)$, $\beta = \frac{1}{4}(b_1-ib_2)$, $\gamma = \frac{1}{4}c$, and

$$\left[\frac{\partial}{\partial z} + \beta(z,\zeta^*)\right] A(z,\zeta^*;\zeta,\zeta^*) = 0 \tag{1.4}$$

$$\left[\frac{\partial}{\partial z^*} + \alpha(\zeta, z^*)\right] A(\zeta, z^*; \zeta, \zeta^*) = 0 \qquad (1.$$

$$A(\zeta, \zeta^*, \zeta, \zeta^*) = 1. \qquad (1.$$

$A(z, z^*, \zeta, \zeta^*)$ is the Riemann function for $\underset{\sim}{L}[A] = 0$ (See Exerci 3.1 of these notes, [20], [21], [24], [39]). Now let D be a simply connected domain with smooth boundary ∂D. Then

$$u(\underset{\sim}{x}) = - \int_{\partial D} B\,[u(\underset{\sim}{\xi}),\ S_M(\underset{\sim}{\xi}; \underset{\sim}{x})] \qquad (1.$$

$$B[u, v] = \sum_{i=1}^{n} (-1)^i \left[b_i u v + v\frac{\partial u}{\partial x_i} - u\frac{\partial v}{\partial x_i}\right] dx, \ldots, dx_i, \ldots, dx_n \quad (1.$$

where S_M is fundamental solution of the adjoint equation $M[u] = 0$. Since for $\underset{\sim}{x} \neq \underset{\sim}{\xi}$, S_M is an analytic function of $\underset{\sim}{x}$ and $\underset{\sim}{\xi}$ for $\underset{\sim}{x}$, $\underset{\sim}{\xi} \in D$ (for $n = 2$ S_M is an analytic function of z, z^*, ζ, ζ^* for z, $\zeta \in D$, z^*, $\zeta^* \in D^*$, where $D^* = \{z^* | \bar{z}^* \in D\}$), provided the coefficients of L have the same property, we have

<u>Theorem A</u> ([20]): If $b_i(\underset{\sim}{x})$, $c(\underset{\sim}{x})$ are analytic functions of $x_1, \ldots, x_n$ in D and $u(\underset{\sim}{x}) \in C^2(D)$, then $u(\underset{\sim}{x})$ is an analytic function of $x_1, \ldots, x_n$ in D.

<u>Theorem B</u> ([39]): Let $n = 2$. If $\alpha(z, z^*)$, $\beta(z, z^*)$, $\gamma(z, z^*)$ are analytic in $D \times D^*$ and $u(x_1, x_2) \in C^2(D)$, then $U(z, z^*) = u\left(\frac{z+z^*}{2}, \frac{z-z^*}{2i}\right)$ is analytic in $D \times D^*$.

2. <u>Existence and Uniqueness of Solutions to the Dirichlet an Cauchy Problems</u>.

Let f, $g \in L_2(D)$. $(f, g) = \iint_D fg$.

Definition: u is a weak solution of $Lu = f$ if $(M\phi, u) = (\phi, f)$ for every $\phi \in C_0^\infty(D)$.

Theorem C ([3]): Let $f(\underset{\sim}{x})$ be analytic in D. Then if u is a weak solution of $Lu = f$, u is analytic in D.

Theorem D ([3]): Let $\phi \in C^{o}(\partial D)$. Then the boundary value problem

$$\begin{aligned} Lu &= f \text{ in } D \\ u &= \phi \text{ on } \partial D \end{aligned} \tag{2.1}$$

where $f \in L_2(D)$, has a (weak) solution if and only if

$$\iint_D f\omega = \int_{\partial D} \phi \frac{\partial \omega}{\partial \nu} \tag{2.2}$$

for all solutions ω of

$$\begin{aligned} M\omega &= 0 \quad \text{in } D \\ \omega &= 0 \quad \text{on } \partial D. \end{aligned} \tag{2.3}$$

The set of solutions of (2.3) is finite dimensional.

Corollary ([3]): The orthogonal complement on ∂D of the space of all boundary values of all solutions of $Lu = 0$ is the finite dimensional space spanned by $\frac{\partial \omega}{\partial \nu}$, ω a solution of (2.3).

Theorem E ([20]): Let $L[u] = 0$, $u \in C^2(D) \cap C^o(\bar{D})$, $c(\underset{\sim}{x}) \leq 0$. Then $u(\underset{\sim}{x})$ achieves its maximum and minimum on ∂D.

Let $\alpha = (\alpha_1, \ldots, \alpha_n)$; $\alpha_j \geq 0$, $|\alpha| = \sum_{j=1}^{n} \alpha_j$,

$D_j = \frac{\partial}{\partial z_j}$, $D^\alpha = D_1^{\alpha 1}, \ldots, D_n^{\alpha}n$, $z^\alpha = z_1^{\alpha 1}, \ldots, z_n^{\alpha}n$.

Theorem F ([28]): Consider

$$D^\beta u = \sum_{|\alpha| \leq |\beta|} a_\alpha D^\alpha u + f \tag{2.4}$$

where f, a_α are analytic functions of $z = (z_1, \ldots, z_n)$ in a neighbourhood of the origin in C^n and prescribe

$$D_j^k (u-\phi) = 0 \text{ when } z_j = 0 \text{ if } 0 \leq k < \beta_j;\ j = 1,\dots,n, \quad (2.$$

where ϕ is analytic in a neighbourhood of the origin. Let A be the set of multi-indices in the sum on the right hand side of (2.4) such that $a_\alpha \not\equiv 0$, and assume that β does not belong to the convex hull of A considered as a subset of R^n. Then there exists a unique analytic solution of (2.4), (2.5) analytic in a neighbourhood Ω of the origin. u depends continuous on the initial data $D_j^k \phi$ in $\Omega' \cap \{z_j = 0\}$ where $\Omega' \subset \Omega$ and the size of Ω' depends only on $\sum_{|\alpha| \leq \beta} |a_\alpha|$.

Remark: It follows from the results of J. Persson, J. Persson, Linear Goursat problems for entire functions when the coefficients are variable, Ann.Scoula Norm. Sup. Pisa (3) 23 (1969), 87-98, that if a_β are constant and ϕ is entire then u is also entire.

II Improperly posed initial value problems

3. Cauchy's Problem in Two Independent Variables

The elliptic Cauchy problem is of interest for the following reasons:

1. analytic continuation

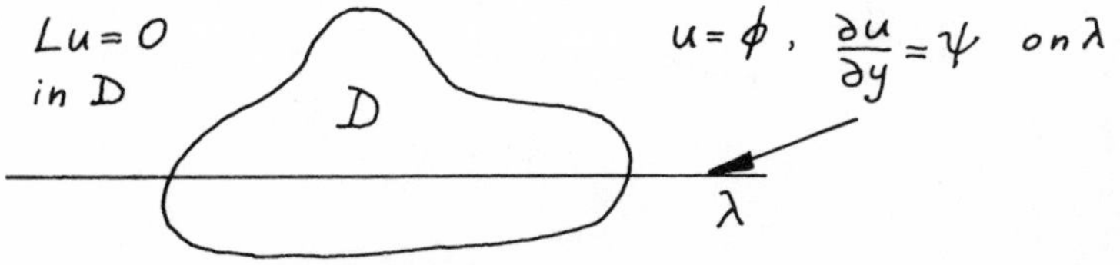

figure 3.1

What is the domain of regularity of u in terms of the domain of regularity of ϕ, ψ and the coefficients of L? Theorem F does not answer this question.

2. inverse boundary value problems

Consider the flow of an incompressible, irrotational fluid about a curved plate B such that behind B there is a "dead water" region Ω.

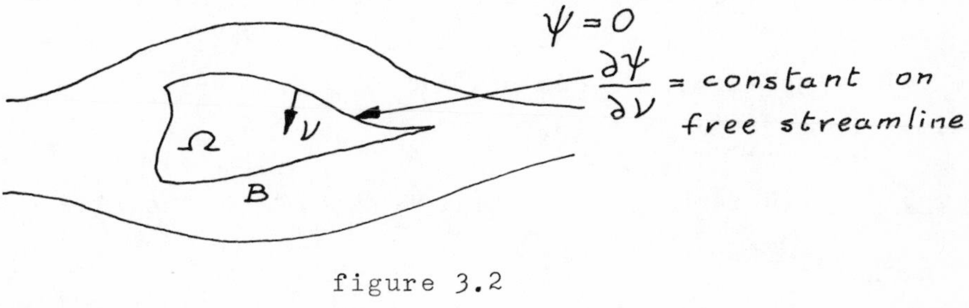

figure 3.2

Let ψ be the stream function. Along the free streamline $\psi=0$. Since the pressure is the same on both sides of the streamline, we have by Bernoulli's equation that $\frac{\partial\psi}{\partial\nu}$ = constant on the streamline. The inverse problem is to find the shape of B given the Cauchy data on the streamline. Analogous problems arise in semilinear and quasilinear equations (see [20]).

3. <u>boundary value problems with incomplete data</u>:

Suppose we have a clamped membrane vibrating with the frequency ω and the slope of deflection is measured on a portion of the boundary.

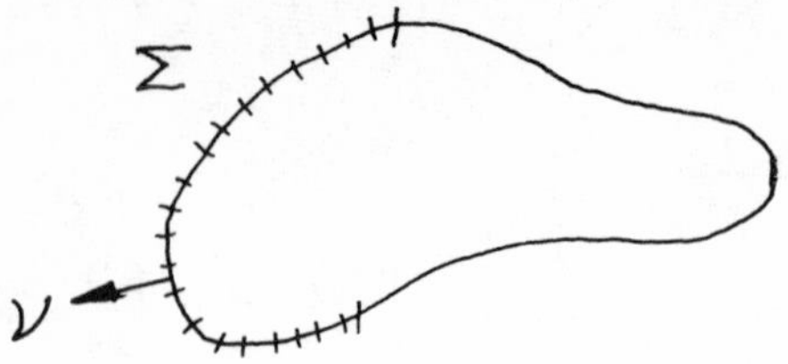

figure 3.3

Then we have the following Cauchy problem:

$$\Delta_2 u + \frac{\omega^2}{c^2} u = 0$$

$$u = 0 \text{ on } \Sigma$$

$$\frac{\partial u}{\partial \nu} = f \text{ on } \Sigma$$

where c = velocity of sound. For a discussion on problems of this type and their solution see [34].

In (2) and (3) the problem is to construct an approximation to the elliptic Cauchy problem, including error estimates. The following example shows why difficulties arise when an attempt is made to do this.

Example 3.1 (Hadamard). Let $u(x,y)$ satisfy the equation

$$\Delta_2 u = 0 \tag{3.1}$$

and the Cauchy data

$$u(x,0) = 0 \qquad u_y(x,0) = \frac{1}{n} \sin nx. \tag{3.2}$$

Then

$$u(x,y) = \frac{1}{n^2} \sin nx \sinh ny. \tag{3.3}$$

is the unique solution to (3.1), (3.2). However as $n \to \infty$ the Cauchy data tends to zero but the solution does not tend to the corresponding zero solution, i.e. Cauchy's problem for elliptic equations does not depend continuously on the initial data.

We now consider the following elliptic Cauchy problem:

$$u_{xx} + u_{yy} = g(x,y,u,u_x,u_y) \tag{3.4}$$

$$u(x,y) = \Phi(x + iy), \; x + iy \in C$$

$$\frac{\partial u(x,t)}{\partial \nu} = \Omega(x + iy), \; x + iy \in C \tag{3.5}$$

where C is a givenanalytic arc. (Regularity conditions on g, Φ, Ω will be prescribed shortly). By the use of a conformal mapping we can assume without loss of generality, that C is a segment of the x-axis containing the origin, i.e. $y = 0$ in (3.5). Setting

$$\begin{aligned} z &= x + iy \\ z^* &= x - iy \end{aligned} \tag{3.6}$$

(3.4), (3.5) becomes

$$U_{zz^*} = f(z,z^*,U,U_z,U_{z^*}) \tag{3.7}$$

where

$$U\left(\frac{z + z^*}{2}, \frac{z - z^*}{2i}\right) = U(z,z^*),$$

and $U(z,z^*) = \Phi(z)$ on $z = z^*$

$$\frac{\partial U(z,z^*)}{\partial z} - \frac{\partial U(z,z^*)}{\partial z^*} = -i\Omega(z) \text{ on } z = z^*. \tag{3.8}$$

Assume i) $f(z,z^*,\xi_1,\xi_2,\xi_3)$ is holomorphic in $G \times G^* \times B_3$ where $G^* = \{z | \bar{z} \in G\}$ and B_3 is a ball containing the origin in ξ_1,ξ_2,ξ_3 space.

ii) G is a disk containing the origin, in particular $G = G^*$.

iii) $\Phi(z)$, $\Omega(z)$ are holomorphic for all $z \in G$.

Now let $s(z,z^*) = U_{zz^*}(z,z^*)$. Then

$$U(z,z^*) = \int_0^z \int_0^{z^*} s(\xi,\xi^*)d\xi^* d\xi + \int_0^z \phi(\xi)d\xi + \int_0^{z^*} \psi(\xi^*)d\xi^* + U(0,0 \quad (3.9$$

$$U_z(z,z^*) = \int_0^{z^*} s(z,\xi^*)d\xi^* + \phi(z) \quad (3.10$$

$$U_{z^*}(z,z^*) = \int_0^z s(\xi,z^*)d\xi + \psi(z^*) \quad (3.1$$

where $\phi(z) = U_z(z,0)$, $\psi(z^*) = U_{z^*}(0,z^*)$.

The initial conditions (3.8) become

$$\int_0^z \int_0^z s\ (\xi,\xi^*)d\xi^* d\xi + \int_0^z \phi(\xi)d\xi + \int_0^z \psi(\xi^*)d\xi^* + U(0,0) = \Phi(z$$

or differentiating with respect to z

$$\int_0^z s(z,\xi^*)d\xi^* + \int_0^z s(\xi,z)d\xi + \phi(z) + \psi(z) = \Phi'(z) \quad (3.1$$

and

$$\int_0^z s(z,\xi^*)d\xi^* + \phi(z) - \int_0^z s(\xi,z)d\xi - \psi(z) = -i\Omega(z). \quad (3.1$$

Equations (3.12), (3.13) imply that

$$\phi(z) = \tfrac{1}{2}[\Phi'(z) - i\Omega(z)] - \int_0 s(z,\xi^*)d\xi^* \quad (3.1$$

$$\psi(z) = \tfrac{1}{2}[\Phi'(z) + i\Omega(z)] - \int_0^z s(\xi,z)d\xi. \quad (3.1$$

Hence, defining the operators B_i, $i = 1,2,3$, by the right hand side of (3.9), (3.10), (3.11) respectively, where $\phi(z)$, $\psi(z)$ are defined by (3.14), and (3.15) (note that $U(0,0) = \Phi(0)$)

Let w_0 be a solution of $Mw_0 = 0$ in ABB'A' of figure 5.1 which has zero Cauchy data on F. Let C be the boundary of the whole rectangle in figure 5.2. Define $w_0 = 0$ outside ABB'A' and consider w_0 in the boundary strip consisting of those points inside C which lie outside the octagon AA'B'BGHKL offigure 5.3. Since we have already shown that solutions with zero Cauchy data on a closed surface are identically zero, $w_0 = 0$ in ABB'A' and the proof is complete . (Note that w_0 has continuous first derivatives across AB which implies that w_0 is a weak solution of $Mw_0 = 0$ and hence w_0 is a strong solution of $Mw_0 = 0$ in the region under consideration).

Exercise 5.2. Let $L[u] = 0$ be a linear second order elliptic equation with analytic coefficients and Laplacian as its principal part. Suppose u has zero Cauchy data on a smooth hypersurface T. Show that $u \equiv 0$ in its domain of definition and conclude that solutions of $L[u] = 0$ have the Runge Approximation Property.

6. The Non-Characteristic Cauchy Problem for Parabolic Equations.

The Stefan problem for the heat equation is defined as follows:

Find $u(x,t)$ and $s(t)$ satisfying

$$u_{xx} - u_t = 0 , \qquad 0 < x < s(t) , \quad 0 < t ,$$

$$u(s(t),t) = 0 , \ u_x(s(t),t) = -\dot{s}(t), \quad 0 < t ,$$

$$u(x,0) = \phi(x) \geq 0, \ 0 \leq x \leq s(0) = b$$

with either $u(0,t) \geq 0$ or $u_x(0,t) \leq 0$ also given. u may be thought of as the temperature distribution in the water component of a one-dimensional ice and water system. The free boundary s(t) represents the interface between the ice and water. The initial temperature ϕ and interface position b are given, along with either the temperature u or its gradient u_x at $x = 0$.

Conversely, we can also consider the inverse Stefan problem, i.e. assume $x = s(t)$ is known and find u, i.e. how must one heat the water in order to melt the ice along a prescribed curve? This is a non-characteristic Cauchy problem for the heat equation. The following example shows that this problem is "improperly posed".

Example 6.1 ([35]):

Let

$$u_n(x,t) = \frac{1}{n^{2k}} \left[e^{nx} \sin(2n^2t+nx) + e^{-nx}\sin(2n^2t+nx) \right].$$

$u_n(x,t)$ satisfies the heat equation and Cauchy data

$$u_n(0,t) = \frac{2}{n^{2k}} \sin 2n^2t, \; u_{nx}(0,t) = 0.$$

$u_n(0,t)$ and its derivatives up to order $k - 1$ tend to zerc as $n \to \infty$ while for $|x| \geq \delta > 0$ $u_n(x,t)$ assumes arbitrari large values as $n \to \infty$, i.e. the inverse Stefan problem is improperly posed. Now consider the Cauchy problem

$$\mathcal{L}[u] \equiv u_{xx} + a(x)u_x + b(x)u - c(x)u_t = F(x,t) \qquad (6$$

$$u(s(t),t) = f(t)$$

$$u_x(s(t),t) = g(t) \qquad (6$$

where a, b, c, F and s are analytic in a sufficiently large neighbourhood of the origin and $x = s(t)$ is non-characteristi We will first construct a fundamental solution S to equation (6.1) which has an essential singularity at $t = \tau$ (as opposed to the usual multi-valued fundamental solution) and use this solve (6.1), (6.2). The results which follow are due to Hill ([26]).

Example 6.2. When $\mathcal{L}[u] \equiv u_{xx} - u_t = 0$ the fundamental soluti S that we will construct is defined by

$$S(x,t;\xi,\tau) = \frac{\sqrt{\pi}(x-\xi)}{2(\tau-t)} \; E \left\{ - \frac{(\xi-x)^2}{4(\tau-t)} \right\}$$

where

$$E(z) = \sum_{n=0}^{\infty} \frac{z^n}{\Gamma(n+3/2)} ,$$

as opposed to the usual fundamental solution

$$w(x,t;\xi,\tau) = \frac{1}{2\sqrt{\pi(\tau-t)}} \exp\left\{-\frac{(\xi-x)^2}{4(\tau-t)}\right\}.$$

The fundamental solution $S(x,t;\xi,\tau)$ of equation (6.1) is a solution of the adjoint equation

$$\begin{aligned} \mathfrak{M}[v] &\equiv v_{xx} - (av)_x + bv + cv_t \\ &\equiv M[v] + cv_t = 0 \end{aligned} \tag{6.3}$$

with initial conditions

$$S(\xi,t;\xi,\tau) = 0, \quad S_x(\xi,t;\xi,\tau) = \frac{-1}{t-\tau}. \tag{6.4}$$

Let

$$S(x,t;\xi,\tau) = \sum_{j=0}^{\infty} S_j(x,\xi) \frac{j!}{(t-\tau)^{j+1}} . \tag{6.5}$$

Equations (6.3) and (6.4) imply that

$$\begin{aligned} &S_0(\xi,\xi) = 0, \quad S_{0x}(\xi,\xi) = -1 \\ &S_j(\xi,\xi) = S_{jx}(\xi,\xi) = 0, \quad j = 1,2,\ldots . \end{aligned} \tag{6.6}$$

Inserting (6.5) into (6.3) gives

$$\mathfrak{M}[S] = \frac{M[S_0]}{(t-\tau)} + \sum_{j=1}^{\infty} \{M[S_j] - cS_{j-1}\} \frac{j!}{(t-\tau)^{j+1}}$$

which implies that

$$\begin{aligned} &M[S_0] = 0 \\ &M[S_j] = cS_{j-1}, \; j = 1,2,\ldots , \end{aligned} \tag{6.7}$$

Equations (6.6) and (6.7) determine the S_j's uniquely. We must now show that (6.5) converges. By Duhamel's principle

$$S_j(x,\xi) = \int_\xi^x R(x,\eta)c(\eta)S_{j-1}(\eta,\xi)d\eta \; ; \; j = 1,2,\ldots \qquad (6.$$

where $M[R] = 0$

$$R(\eta,\eta) = 0 \qquad R_x(\eta,\eta) = 1.$$

On any compact interval $|\xi|, |\eta|, |x| \leq h$ from the analytic theory of ordinary differential equations there exist constan M_0, K, C such that

$$|S_{0x}(x,\xi)| \leq M_0, \quad |S_0(x,\xi)| \leq M_0|x-\xi|$$

$$|R_x(x,\eta)| \leq K, \quad |R(x,\eta)| \leq K\,|x-\eta|$$

$$|c(\eta)| \leq C.$$

From the observation that

$$\left| \int_\xi^x |x-\eta| \frac{|\eta-\xi|^{2j-1}}{(2j-1)!} d\eta \right| = \frac{|x-\xi|^{2j+1}}{(2j+1)!}$$

we have by induction on (6.8) that

$$|S_j(x,\xi)| \leq M_0 M^j \frac{|x-\xi|^{2j+1}}{(2j+1)!}$$

$$(6.$$

$$|S_{jx}(x,\xi)| \leq M_0 M^j \frac{|x-\xi|^{2j}}{(2j)!}$$

where $M = KC$. Equation (6.9) implies that the series (6.5) co verges absolutely and uniformly and can be differentiated termwise.

Now consider the identity

$$\int\int_D \{v\mathcal{L}[u] - u\mathcal{M}[v]\}dxdt \qquad (6.1$$

$$= \int_{\partial D} \{(vu_x - uv_x + auv)dt + cuvdx\}$$

where D is some two dimensional chain in the region of analyticity of the integrand and ∂D is its one dimensional boundary.

In equation (6.10) let $v = S$, u be a solution of $\mathcal{L}[u] = F$, and let D be the lateral surface of a cylinder that wraps around $t = \tau$ and has γ_0 and γ as its two rims, where γ_0 is a loop about $t = \tau$ in the plane $x = \xi$ and γ is some other loop about $t = \tau$ (see figure 6.1 below).

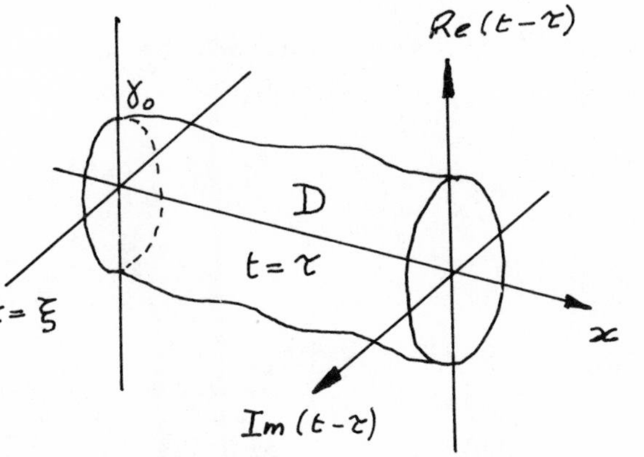

figure 6.1

Because of (6.4) we have

$$\oint_{\gamma 0} \{(Su_x - uS_x + auS)dt + cuSdx\} = \oint_{\gamma_0} \frac{u(\xi,t)}{t-\tau}dt$$
$$= 2\pi i \ u(\xi,t)$$

and hence (6.10) becomes

$$u(\xi,\tau) = \frac{1}{2\pi i}\oint_\gamma (Su_x - uS_x + auS)dt + cuSdx + \frac{1}{2\pi i}\int_D\int SFdxdt$$

Placing the cycle γ on the two dimensional manifold $x = s(t)$ (t complex) where the Cauchy data (6.2) is prescribed gives

$$u(\xi,\tau) = \frac{1}{2\pi i}\oint_\gamma \left\{(Su_x - uS_x + auS + cuS\dot{s} + \oint_\xi^{s(t)} SFdx\right\}dt, \tag{6.11}$$

the desired solution of the Cauchy problem (6.1), (6.2).

Suppose now $F \equiv 0$ and $s(t) = \text{constant} = x_0$. Then (6.11) becomes

$$u(\xi,\tau) = \operatorname*{Res}_{t=\tau} \left\{ Su_x - uS_x + auS + cuS\dot{s} \right\}$$

$$= \sum_{j=0}^{\infty} \frac{\partial^j}{\partial\tau^j} \{S_j(x_0,\xi)u_x(x_0,\tau) \tag{6.$$

$$+ [a(x_0)S_j(x_0,\xi) - S_{jx}(x_0,\xi)]u(x_0,\tau)\}.$$

Suppose the Cauchy data, as a function of τ, is analytic for $|\tau| \leq \rho$. Then by Cauchy's inequality there exists a constant A such that

$$\left|\frac{\partial^j}{\partial\tau^j} u(x_0,\tau)\right| \leq A\frac{j!}{\rho^j} \tag{6.$$

$$\left|\frac{\partial^j}{\partial\tau^j} u_x(x_0,\tau)\right| \leq A\frac{j!}{\rho^j} \quad .$$

The estimates (6.9) and (6.13) imply that the series (6.12) is dominated (up to multiplication by a constant) by

$$\sum_{j=0}^{\infty} \frac{1}{j!} \left\{ \frac{M|x_0-\xi|^2}{\rho} \right\}^j = \exp \left\{\frac{M|x_0-\xi|_2}{\rho}\right\}, \tag{6.$$

which implies the following theorem:

<u>Theorem 6.1</u> ([26]): Assume the coefficients of $\mathcal{L}[u] = 0$ a entire functions of x and let u be a solution of $\mathcal{L}[u] = 0$ which is a real analytic function of x and t in the circle $x^2 + t^2 < \rho^2$. Then u can be continued as an analytic function of x and t into the strip $-\rho < t < \rho$, $-\infty < x <$

<u>Remarks</u>: For related results for parabolic equations in two space variables see [27].

7. <u>Improperly Posed Initial-Value Problems for Hyperbolic Equations</u>.

Consider the equation

$$u_{x_1x_1} = u_{x_2x_2} + u_{x_3x_3} + q(x_1,x_2,x_3)u - f(x_1,x_2,x_3) \qquad (7.1)$$

nd assume q and f are entire functions of their independent complex) variables. The Cauchy problem along a space-like urface is well posed:

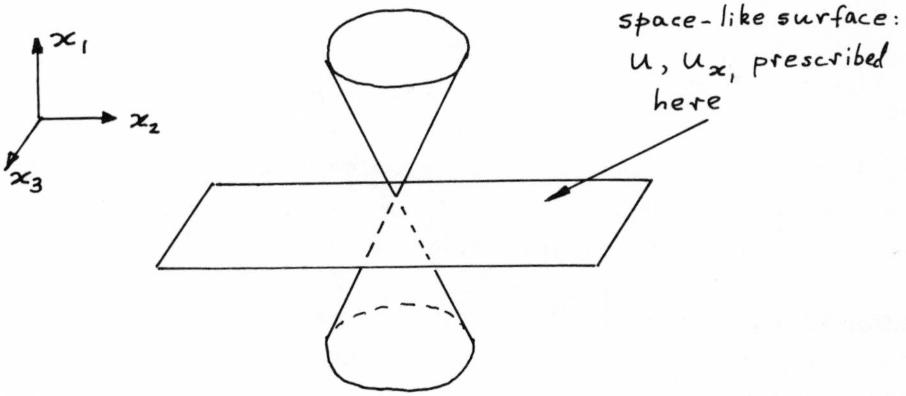

figure 7.1

owever the Cauchy problem along a time-like surface is mproperly posed (c.f. [20], p. 176):

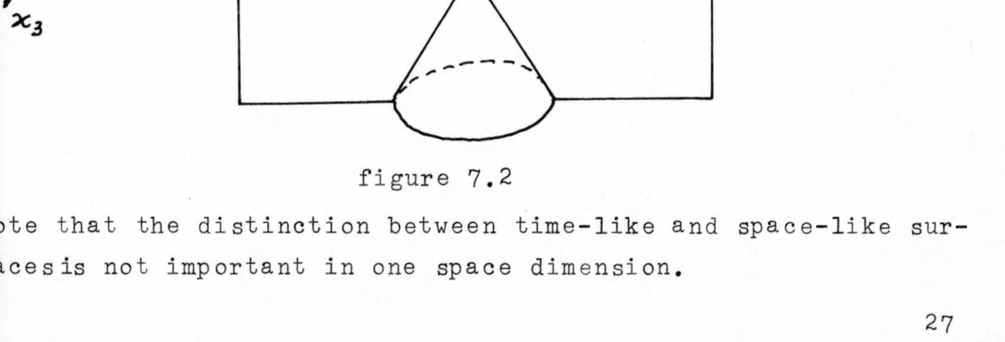

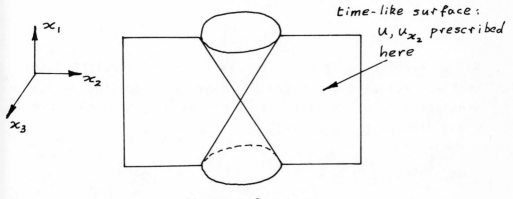

figure 7.2

ote that the distinction between time-like and space-like suracesis not important in one space dimension.

Example 7.1:

$$u = \frac{1}{n^2} \sin hnx_2 \sin nx_3$$

is a solution of

$$u_{x_1x_1} = u_{x_2x_2} + u_{x_3x_3}$$

satisfying

$$u(x_1,0,x_3) = 0 \quad u_{x_2}(x_1,0,x_3) = \frac{1}{n} \sin nx_3.$$

As $n \to \infty$ the initial data tends to zero but the solution does not, i.e. the Cauchy problem for hyperbolic equations along a time-like surface is improperly posed.

Example 7.2: Suppose we have a clamped, vibrating membrane and the slope of the deflection is measured on portion Σ of the boundary.

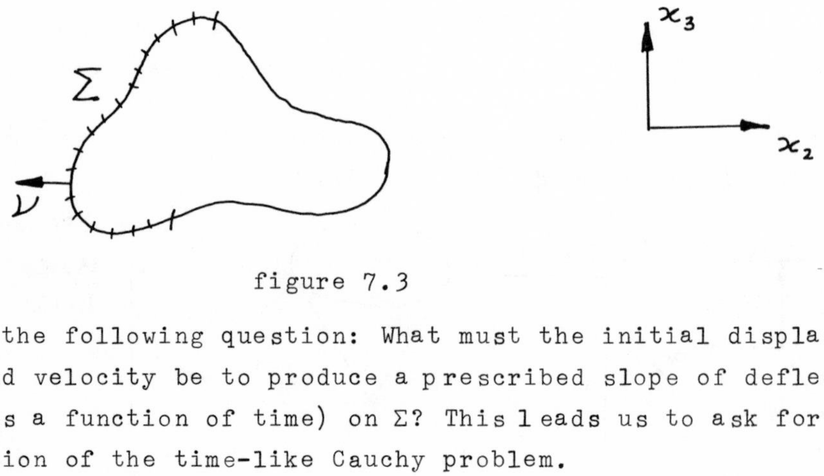

figure 7.3

We ask the following question: What must the initial displacement and velocity be to produce a prescribed slope of deflection (as a function of time) on Σ? This leads us to ask for a solution of the time-like Cauchy problem.

$$\frac{1}{c^2} u_{x_1x_1} = u_{x_2x_2} + u_{x_3x_3}$$

$$u = 0 \text{ on } \Sigma$$

$$\frac{\partial u}{\partial \nu} = g \text{ on } \Sigma$$

(where c = velocity of sound) and to then evaluate u and $\frac{\partial u}{\partial x_1}$ at $x_1 = 0$. Suppose the plane $x_2 = 0$ in figure 7.2 is bent 45° on each side of the x_3 axis to form two intersecting planes tangent to a genertor of each nape of the characteristic cone. The exterior characteristic initial value problem for equation (7.1) is to find a solution u of (7.1) in the quarter space bounded by these intersecting planes such that u assumes prescribed values on each of the two planes. This problem is also improperly posed ([17]).

Example 7.3 ([17]): Suppose u is a solution of the three dimensional wave equation in spherical coordinates

$$\frac{\partial^2 u}{\partial t^2} = \nabla^2 u = \frac{\partial^2 u}{\partial r^2} + \frac{2}{r}\frac{\partial u}{\partial r} + \frac{1}{r^2 \sin\theta}\frac{\partial}{\partial\theta}\left(\sin\theta\frac{\partial u}{\partial\theta}\right) + \frac{1}{r^2\sin^2\theta}\frac{\partial^2 u}{\partial\phi^2} \tag{7.2}$$

that is defined in $r \geq a > 0$ and vanishes for $t \leq r$. (This represents an outgoing wave produced by sources in $r < a$). It can be shown that if $t - r = \tau$ is bounded then

$$\lim_{r\to\infty} \{ru(r, \theta, \phi, r + \tau)\} = f(\theta, \phi, \tau)$$

exists i.e. $ru \sim f(\theta, \phi, t - r)$ for large r. The inverse problem is, given the "radiation field" f, to determine, u. Set

$$ru = v$$

$$t - r = \tau$$

$$t + r = \frac{1}{\sigma}$$

Then (7.2) becomes

$$\frac{\partial^2 v}{\partial\tau\partial\sigma} + \frac{1}{(1-\sigma\tau)^2}\left\{\frac{1}{\sin\theta}\frac{\partial}{\partial\theta}\left(\sin\theta\frac{\partial v}{\partial\theta}\right) + \frac{1}{\sin^2\theta}\frac{\partial^2 v}{\partial\phi^2}\right\} = 0 \tag{7.3}$$

and the data for the inverse problem is

$$v\Big|_{\tau = 0} = 0, \; 0 \leq \sigma \leq \frac{1}{2a}$$

$$v\Big|_{\sigma = 0} = f(\theta, \phi, \tau), \; \tau \geq 0 \qquad (7.$$

$$f(\theta, \phi, 0) = 0.$$

Equations (7.3) and (7.4) constitute an exterior characterist initial value problem.

We will need the following definition:

Definition 7.1: A function $g(x_1,x_2)$ of two real variables x and x_2 will be said to be partially analytic with respect to x_1 for $x_1 = a$ in the interval $\alpha \leq x_2 \leq \beta$ provided it can be represented by a series of the form

$$g(x_1,x_2) = b_0(x_2) + b_1(x_2)(x_1 - a) + b_2(x_2)(x_1 - a)^2 + \ldots \qquad (7.$$

whose coefficients are continuous functions of x_2 in the int val $\alpha \leq x_2 \leq \beta$ and provided that the series (7.5) converges absolutely and uniformly for $\alpha \leq x_2 \leq \beta$, $|x_1 - a| \leq \gamma$. The region $\alpha \leq x_2 \leq \beta$, $|x_1 - a| \leq \gamma$ is known as the region of partial analyticity. The extension to more variables is evid First consider the time-like Cauchy problem. Let

$$x = x_3 - x_1$$

$$y = x_1 + x_3 \qquad (7.$$

$$z = x_2.$$

Then (7.1) becomes

$$L[u] \equiv u_{zz} + 4u_{xy} + Q(x,y,z)u = F(x,y,z) \qquad (7$$

where $F(x,y,z) = f(x_1,x_2,x_3), Q(x,y,z) = q(x_1,x_2,x_3)$. Let u and v be "well behaved" functions to be prescribed shortly a integrate the identity

$$vL[u] - uL[v] = (2vu_y - 2v_y u)_x + (2vu_x - 2v_x u)_y + (vu_z - uv_z)_z \qquad (7.8)$$

ver the torus $D \times \Omega$ where $\Omega = \Omega(\zeta)$: $|z - \zeta| = \delta > 0$ is a ircle in the complex z plane and $D = D(\zeta) \subset \mathbb{R}^2$ ($\mathbb{R}^2$ is the uclidean plane) is as in figure (7.4) below.

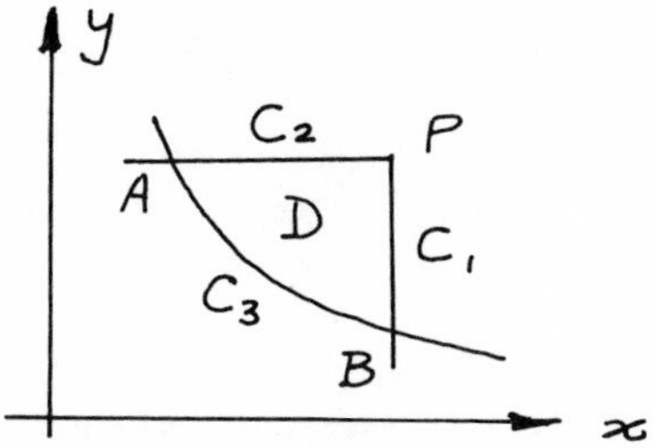

figure 7.4

$_3 = C_3(\zeta)$ is on the complex extension with respect to ζ of he intersection of the plane $z = \zeta$ with the smooth convex nitial surface on which the Cauchy data is prescribed. It s assumed that the normal to this surface is never parallel o the z-axis and that the initial surface is partially nalytic with respect to z. Note that in the special case hen C_3 is independent of ζ the cylinder C_3 is time-like in uclidean three space $\mathbb{R}^3$, but the cylinder C_3' in figure (7.5) elow is space like.

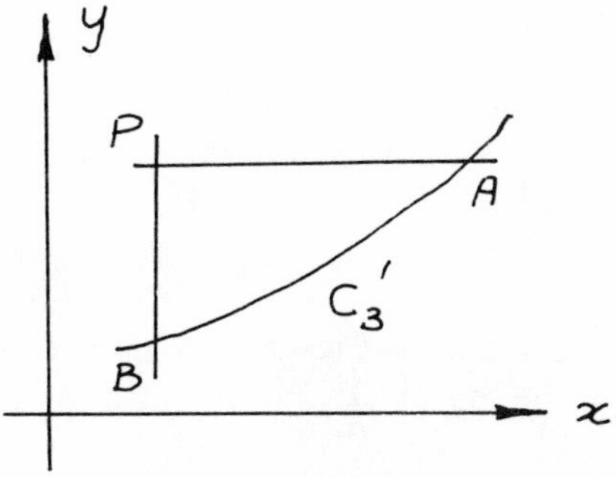

figure 7.5

Integrating (7.8) over $D \times \Omega$ gives

$$\iiint\limits_{D\times\Omega} (vL[u] - uL[v])\,dxdydz$$

$$+ \int_{\Omega} [2v(A,z)u(A,z) + 2v(B,z)u(B,z) + 4v(P,z)u(P,z)]dz$$

$$+ 4 \iint\limits_{C_1\times\Omega} uv_y\,dydz - 4 \iint\limits_{C_2\times\Omega} uv_x\,dxdz \qquad (7.9$$

$$+ 2 \iint\limits_{C_3\times\Omega} [(uv_y - vu_y)dydz - (uv_x - vu_x)dxdz] = 0$$

(Note that $dxdy = 0$ on $\partial D \times \Omega$). In equation (7.9) $v(A,z)=v(x,$ where $A = (x,y)$, etc.. Now, let

1) u be a C^2 solution of $L[u] = F$ such that u and its der vatives of order less than or equal to two are partial analytic with respect to z in some neighbourhood of a smooth (time-like) convex surface, where $C_3 = C_3(\zeta)$ lies on the complex extension of the intersection of this surface with the plane $z = \zeta$.

2) v be a fundamental solution of $L[u] = 0$ such that

$$v_y = 0 \quad \text{on } C_1 \times \Omega \qquad (7.1$$

$$v_x = 0 \quad \text{on } C_2 \times \Omega \qquad (7.1$$

and at the point $(P,z) = (\xi,\eta,z)$

$$v(P,z) = \frac{1}{8\pi i(z-\zeta)} + \text{analytic function of } (z-\zeta) \qquad (7.1$$

We must now construct v. A fundamental solution S of $L[v] = 0$ is of the (normalized) form

$$S = \frac{1}{8\pi iR} + \sum_{\ell=1}^{\infty} U_\ell R^{2\ell-1} + W \qquad (7.1$$

here $R = \sqrt{(z-\zeta)^2 + (x-\xi)(y-\eta)}$. The $U_\ell \equiv U_\ell(x,y,z;\ \xi,\eta,\zeta)$ can e computed recursively and are entire functions of their ndependent variables (since Q is - c.f. [20]). W is a regular olution of $L[w] = 0$. Let W satisfy the boundary conditions

$$W = -\sum_{=1}^{\infty} U_\ell (z-\zeta)^{2\ell-1} \quad \text{on } x = \xi \tag{7.14}$$

$$W = -\sum_{=1}^{\infty} U_\ell (z-\zeta)^{2\ell-1} \quad \text{on } y = \eta. \tag{7.15}$$

exists from Theorem F and is entire since Q is. Then S def- ned by (7.13) satisfies equations (7.10)-(7.12) and we can set $= S$ in (7.9) provided $|z - \zeta|^2 > |(x-\xi)(y-\eta)|$ i.e. z lies out- ide the cut in the complex z plane along a line parallel to he imaginary axis between $\zeta \pm i\sqrt{(x-\xi)(y-\eta)}$.

Note that in view of equations (7.10)-(7.12) W can actually e chosen in a variety of different ways, e.g. when $=$ constant $= \lambda^2$ a possible choice for $v = S_\lambda$ is

$$S_\lambda = \frac{\cos \lambda R}{8\pi\, iR} \, . \tag{7.16}$$

rom (7.12) we have

$$4 \int_\Omega v(P,z)u(P,z)dz = u(\xi,\eta,\zeta) \tag{7.17}$$

nd hence (7.9) becomes (setting $v = S$)

$$\begin{aligned} u(\xi,\eta,\zeta) = &-2\int_{\Omega(\zeta)} [S(A,z;\xi,\eta,\zeta)u(A,z) + S(B,z;\xi,\eta,\zeta)u(B,z)]dz \\ &+2\int\int_{C_3(\zeta)\times\Omega(\zeta)} [u(x,y,z)S_x(x,y,z;\xi,\eta,\zeta) - S(x,y,z;\xi,\eta,\zeta)u_x(x,y,z)]dxdz \\ &-2\int\int_{C_3(\zeta)\times\Omega(\zeta)} [u(x,y,z)S_y(x,y,z;\xi,\eta,\zeta) - S(x,y,z;\xi,\eta,\zeta)u_y(x,y,z)]dydz \\ &-\iiint_{D(\zeta)\times\Omega(\zeta)} S(x,y,z;\xi,\eta,\zeta)F(x,y,z)dxdydz. \end{aligned} \tag{7.18}$$

Equation (7.18) gives the solution of the time-like Cauchy problem along a smooth convex surface.

Note that

1) Partial analyticity of the Cauchy data and its derivatives of order less than or equal to two along $y = y(x)$ implies that u and its derivative of order less than or equal to two are partially analytic in $D \times \Omega$.

2) Recall that S is an analytic function of z outside the cut between $\zeta \pm i\sqrt{(x-\xi)(y-\eta)}$. Let $G(x,y)$ be an arbitrarily small neighbourhood of this cut. Then (by deforming Ω) equation (7.18) shows that at the point (ξ, η, ζ) u depends continuously on its Cauchy data in $C_3 \times G$, where $G \supset G(x,y)$ for all points $(x,y) \in C_3$.

By deforming the curve C_3 onto the characteristics $C_4 = AT$, $C_5 = TB$ (see figure 7.6 below), and integrating by parts to eliminate the partial derivatives of u along these characteristics, we arrive at the solution of the exterior characteristic initial-value problem ([7]):

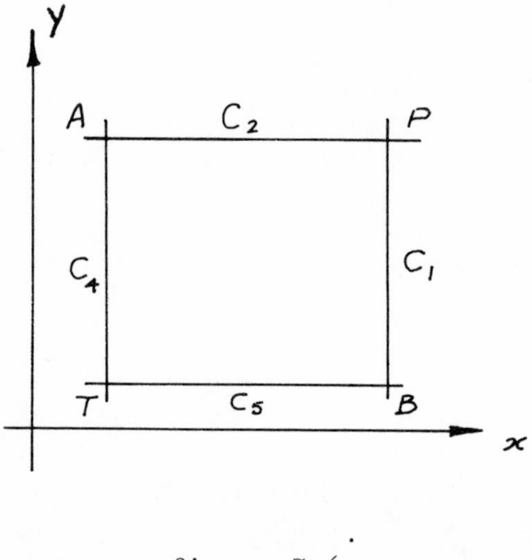

figure 7.6

$$u(\xi,\eta,\zeta) = -2\int_{\Omega}[S(A,z;\xi,\eta,\zeta)u(A,z) + S(B,z;\xi,\eta,\zeta)u(B,z) - S(T,z;\xi,\eta,\zeta)u(T,z)]dz$$

$$-4\int\int_{C_4\times\Omega} S_y(x,y,z;\xi,\eta,\zeta)u(x,y,z)dydz \tag{7.19}$$

$$+4\int\int_{C_5\times\Omega} S_x(x,y,z;\xi,\eta,\zeta)u(x,y,z)dxdz$$

$$-\int\int\int_{D\times\Omega} S(x,y,z;\xi,\eta,\zeta)F(x,y,z)dxdydz.$$

Note that the regions of integration are now independent of ζ.
Remark: Similar integral representations can be obtained for the Cauchy and Goursat problems for the elliptic equation

$$u_{x_1x_1} + u_{x_2x_2} + u_{x_3x_3} + q(x_1,x_2,x_3)\mu = f(x_1,x_2,x_3) \tag{7.20}$$

with data along a convex analytic surface. To see this set

$$X = x_1$$
$$Z = x_2 + ix_3$$
$$Z^* = x_2 - ix_3.$$

Then (7.2) becomes

$$U_{XX} + 4U_{ZZ^*} + Q(X,Z,Z^*)U = F(X,Z,Z^*)$$

which is of the same form as (7.7). Repeating the previous analysis now leads to the representations (7.18) and (7.19) (with z replaced by X, x replaced by Z, and y replaced by Z*). for the solution of the Cauchy and complex Goursat problem respectively.

III Integral operators for elliptic equations

8. Integral Operators in Two Independent Variables

In one sense the integral representations obtained in sections 6 and 7 can be viewed as integral operators for parabolic and hyperbolic equations. Here we obtain integral operators for elliptic equations such that the kernel of the operator is an entire function of its independent variables.

Consider the self-adjoint equation (see, however, the remar at the end of this section).

$$\Delta_2 u - q(x,y)u = 0 \tag{8.1}$$

Where $u(x,y) \in C^2(D)$, D is simply connected with C^2 boundary ∂D, $q(x,y)$ is a real valued (for x, y real) entire function of the (complex) variables x,y, (with minor modifications we could have considered q to be analytic only in some polycylinder). We want to generalize the following example:

Example 8.1: Suppose $\Delta u = 0$ in D. Then $u = \text{Re } f(z)$ where $f(z$ is analytic. But by Runge's theorem $\{z^n\}$ is a complete family of analytic functions in D. Hence $\{\text{Re } z^n\} = \{r^n \cos n\theta\}$ and $\{\text{Im } z^n\} = \{r^n \sin n\theta\}$ together form a complete family of solutions for $\Delta u = 0$ in D. To approximate solutions of $\Delta u = 0$ in D, $u = f$ on ∂D, set $u_N = \sum_{m=0}^{N} a_m r^n \cos n\theta + b_m r^n \sin n\theta$ and minimise $|U_N - f|$ on ∂D. By Theorem E this minimizes $|U_N - U|$ in D. A complete family of solutions can also be used to approximate solutions to Cauchy's problem:

Example 8.2: Suppose $\Delta_n u + q(\underset{\sim}{x})u = 0$ in D and $u = f, \frac{\partial u}{\partial \nu} = g$ on $\Sigma \subset \partial D$. Then if $|u| \leq 2M$ in D then there exist constants

STRAVINSKY

STRAVINSKY

ANDRÉ BOUCOURECHLIEV

translated by Martin Cooper

HM

HOLMES & MEIER

NEW YORK

Published in the
United States of America 1987 by
Holmes & Meier Publishers, Inc.
30 Irving Place
New York, NY 10003

Originally published in French under the title
Stravinsky by Librairie Arthème Fayard, 1982

Library of Congress Cataloging in Publication Data
Boucourechliev, André.
Stravinsky.
Translation of: Stravinsky.
Bibliography: p.
Includes indexes.
1. Stravinsky, Igor, 1882–1971. 2. Composers—
Biography. I. Title.
ML410.S932B6813 1987 780'.92'4 [B] 86–33488
ISBN 0–8419–1058–8
ISBN 0–8419–1162–2 (pbk)

Printed in Great Britain

$_1 = C_1(M)$, $C_2 = C_2(M)$ such that

$$\max_{\underset{\sim}{x} \in \bar{D}} |u|^{2/\delta} \leq C_1 \int_\Sigma |u|^2 ds + C_2 \int_\Sigma |\frac{\partial u}{\partial \nu}| ds$$

here $0 < \delta < 1$. (c.f. [34]).

Let $\{\phi_k\}$ be a complete family of solutions to $\Delta_n u + q(\underset{\sim}{x})u = 0$

$$u_N = \sum_{k=1}^{N} a_k \phi_k(\underset{\sim}{x}) .$$

o approximate u (under the assumption $|u| \leq M$ in D) use the ayleigh Ritz procedure to minimize

$$C_1 \int_\Sigma (f-u_N)^2 ds + C_2 \int_\Sigma (g - \frac{\partial u_N}{\partial \nu})^2 ds$$

ubject to the constraint $|u_N| \leq M$ in D.

ow let

$$z = x + iy \qquad z^* = x - iy \tag{8.2}$$

e a mapping of $C_2 \to C_2$ (C_2 denotes the space of two complex ariables). Equation (8.1) becomes

$$L(U) = \frac{\partial^2 U}{\partial z \partial z^*} + Q(z,z^*)U = 0 \tag{8.3}$$

here $Q(z,z^*) = -\frac{1}{4} q\left(\frac{z+z^*}{2}, \frac{z-z^*}{2i}\right)$

$$U(z,z^*) = u\left(\frac{z+z^*}{2}, \frac{z-z^*}{2i}\right) . \tag{8.4}$$

<u>Theorem 8.1 ([2])</u>: Let $E(z,z^*,t)$ be an analytic function of t, z, z^* for $|t| \leq 1$ and z, z^* in some neighbourhood of the origin which satisfies the partial differential equation

$$-(1 - t^2)E_{z^*t} + \frac{1}{t} E_{z^*} - 2tzL(E) = 0 \tag{8.5}$$

nd is such that E_{z^*}/zt is continuous at $z = 0$, $t = 0$. Then if $f(z)$ is an analytic function of z in a neighbourhood of $z = 0$,

$$\mathcal{U}(z,z^*) = \underset{\sim}{P}_2\{f\} = \int_{-1}^{1} E(z,z^*,t) f(\tfrac{z}{2}(1-t^2)) \frac{dt}{(1-t^2)^{\frac{1}{2}}} \quad (8.$$

will be a solution of (8.3) in a sufficiently small neighbourhood of $z = 0$, $z^* = 0$.

Proof:

$$\mathcal{U}_{zz^*} = \int_{-1}^{1}\left(E_{zz^*} f(\tfrac{z}{2}(1-t^2)) + E_{z^*} \frac{\partial f(\frac{z}{2}(1-t^2))}{\partial z}\right)\frac{dt}{\sqrt{1-t^2}}$$

But $f_z = - f_t(1-t^2)/2zt$ which implies that

$$\mathcal{U}_{zz^*} = \int_{-1}^{1} \left(E_{zz^*} f - E_{z^*}(1-t^2)(2zt)^{-1} f_t\right) \frac{dt}{\sqrt{1-t^2}} .$$

Integrating the second term by parts gives

$$\mathcal{U}_{zz^*} = \int_{-1}^{1} E_{zz^*} f \frac{dt}{\sqrt{1-t^2}} - \left(\frac{E_{z^*}\sqrt{1-t^2}}{2zt} f\left(\frac{z}{2}(1-t^2)\right)\right)_{t=-1}^{t=+1}$$

$$+\int_{-1}^{1} \left(\frac{E_{z^*}\sqrt{1-t^2}}{2zt}\right)_t f dt = \int_{-1}^{1}\left(\frac{E_{zz^*}}{\sqrt{1-t^2}} + \left(\frac{E_{z^*}\sqrt{1-t^2}}{2zt}\right)_t\right) f\left(\frac{z}{2}(1-t^2)\right) d$$

Equations (8.3), (8.5) now imply the theorem.

Theorem 8.2 ([2]): There exists a function $E(z,z^*,t)$ satisfying all the conditions of Theorem 8.1 and also

$$E(0,z^*,t) = E(z,0,t) = 1. \quad (8.$$

If $q(x,y)$ is entire then $E(z,z^*,t)$ is an entire function of z and z^*.

Proof: Let

$$E(z,z^*,t) = 1 + \sum_{n=1}^{\infty} t^{2n} z^n \int_0^{z^*} P^{(2n)}(z,z^*)dz^*. \quad (8.$$

Substitute (8.8) into (8.5) and compare powers of t.

leads to the problem of finding $s(z,z^*)$ satisfying

$$s(z,z^*) = f(z,z^*,B_1[s(z,z^*)],B_2[s(z,z^*)],B_3[s(z,z^*)]).$$

Let HB be the Banach space of functions of two complex variables which are holomorphic and bounded in $G \times G^*$ with norm

$$\| s \|_\lambda = \sup_{(z,z^*)\in G\times G^*} \{e^{-\lambda(|z|+|z^*|)} |s(z,z^*)|\}. \tag{3.17}$$

where $\lambda > 0$ is a fixed constant.

<u>Theorem 3.1</u> ([6]): The operator T defined by

$$Ts = f(z,z, B_1 s, B_2 s, B_3 s)$$

maps a closed ball of HB into itself and is a contraction mapping.

<u>Proof</u>: Since f is holomorphic in a compact subset of the space of five complex variables, by Schwarz's lemma (c.f.[21]) a Lipschitz condition holds there with respect to the last three arguments, i.e.

$$\begin{aligned} &|f(z,z^*,\xi_1,\xi_2,\xi_3) - f(z,z^*,\xi_1^0,\xi_2^0,\xi_3^0)| \\ &\leq C_0\{|\xi_1-\xi_1^0| + |\xi_2-\xi_2^0| + |\xi_3-\xi_3^0|\} \end{aligned} \tag{3.18}$$

where C_0 is a positive constant. Hence for $s_1,s_2 \in HB$ and G sufficiently small

$$\begin{aligned} &\| Ts_1-Ts_2\|_\lambda \leq C_0\{ \|B_1 s_1-B_1 s_2 \|_\lambda + \|B_2 s_1-B_2 s_2 \|_\lambda \\ &+ \|B_3 s_1-B_3 s_2 \|_\lambda\}. \end{aligned} \tag{3.19}$$

From estimates of the form

$$\left|\int_0^z s(\xi, z^*)d\xi\right| \leq \int_0^{|z|} \|s\|_\lambda e^{\lambda|\xi|+\lambda|z^*|}|d\xi|$$

$$\leq \frac{1}{\lambda} e^{\lambda|z|+\lambda|z^*|} \|s\|_\lambda \qquad (3.2$$

i.e.

$$\left\|\int_0^z s(\xi, z^*)d\xi\right\|_\lambda \leq \frac{\|s\|_\lambda}{\lambda} \qquad (3.2$$

we have

$$\|B_i s_1 - B_i s_2\|_\lambda \leq \frac{N_i}{\lambda} \|s_1 - s_2\|_\lambda; \; i = 1,2,3 \qquad (3.2$$

where N_i are positive constants independent of λ. Hence there exists a positive constant M such that

$$\|Ts_1 - Ts_2\|_\lambda \leq \frac{M}{\lambda} \|s_1 - s_2\|_\lambda \qquad (3.2$$

and for every s∈HB

$$\|Ts\|_\lambda \leq \frac{M}{\lambda} \|s\|_\lambda + \|T_0\|_\lambda$$

$$\leq \frac{M}{\lambda} \|s\|_\lambda + \frac{M_0}{2} \qquad (3.2$$

where M_0 is a positive constant. Hence for $\|s\|_\lambda \leq M_0$ an λ sufficiently large, $\|Ts\|_\lambda \leq M_0$ i.e. T takes a closed ball in HB into itself. Equation (3.23) shows that for λ sufficiently large T is a contraction mapping.

<u>Corollary</u> ([6]): There exists a stable iterative procedure fo solving the semilinear elliptic Cauchy problem in two independent variables. When (3.4) is linear global solutions are obtained; if the Cauchy data is analytic in G and the coefficients are analytic functions of z and z* in G x G*, then the solution is an analytic function of z and z* in G x G*.

Exercise 3.1. Use the above method of exponential majorization to construct the complex Riemann function (see Section 1). Compare to [20], pp. 139-141, and [39].

Remark: For related results see [20], pp. 625-631 and [24].

4. Cauchy's Problem for Quasilinear Systems.

Consider a quasilinear system of m partial differential equations of order one in n + 1 independent variables $x,\ldots,x_n, t$

$$\frac{\partial u}{\partial t} = \sum_{j=1}^{n} A_j \frac{\partial u}{\partial x_j} + B \tag{4.1}$$

where

$$u = u(x,t) = \begin{bmatrix} u_1 \\ \vdots \\ u_m \end{bmatrix}$$

and $A_j = A_j(x,t,u)$, $j = 1,\ldots,n$ are $m \times m$ matrices which are analytic functions of $x = (x,\ldots,x_n)$, t and u, and $B = B(x,t,u)$ is a column vector of analytic function of x,t and u. (Any system of partial differential equations can be written as (4.1) if $t = 0$ is not characteristic: c.f. [20], pp. 6-12). Pose the Cauchy problem for (4.1) by prescribing the initial condition

$$u(x,0) = f(x) \tag{4.2}$$

where f(x) is analytic. By the Cauchy-Kowalewski Theorem (c.f. [20]) there exists locally an analytic solution u of (4.1), (4.2).

Now keep t real and replace x_j by $z_j = x_j + iy_j$ (x_j, y_j are real). Then

$$\frac{\partial U}{\partial z_j} = \frac{1}{2}\left(\frac{\partial u}{\partial x_j} - i\frac{\partial u}{\partial y_j}\right) \tag{4.3}$$

where $U(z,\bar{z},t) = u(x,t)$, $z = (z_1,\ldots,z_n)$, $\bar{z} = (\bar{z}_1,\ldots,\bar{z}_n)$.

Furthermore, since U is an analytic function of z_j, $j = 1, \ldots, n$, the n Cauchy-Riemann equations are satisfied:

$$\frac{\partial U}{\partial \bar{z}_j} = \frac{1}{2}\left(\frac{\partial u}{\partial x_j} + i\frac{\partial u}{\partial y_j}\right) = 0. \qquad (4$$

Equations (4.1), (4.2) become

$$\frac{\partial U}{\partial t} = \sum_{j=1}^{n} A_j \frac{\partial U}{\partial z_j} + B \qquad (4$$

$$U(z,0) = f(z). \qquad (4$$

Let A_j^* be the transpose of the complex conjugate of A_j. Multiplying (4.4) by A_j^* and adding it to (4.5) gives (see [20])

$$\frac{\partial U}{\partial t} = \sum_{j=1}^{n} A_j \frac{\partial U}{\partial z_j} + \sum_{j=1}^{n} A_j^* \frac{\partial U}{\partial \bar{z}_j} + B$$

or by (4.3) and (4.4)

$$\frac{\partial u}{\partial t} = \sum_{j=1}^{n} \frac{A_j + A_j^*}{2} \frac{\partial u}{\partial x_j} + \sum_{j=1}^{n} \frac{A_j - A_j^*}{2i} \frac{\partial u}{\partial y_j} + B. \qquad (4$$

Recall that a system of the form (4.1) is symmetric hyperbolic if and only if the A_j are symmetric, i.e. for arbitrary but fixed λ_j the roots λ of the polynomial

$$\det\left|\sum_{j=1}^{n} \lambda_j A_j - \lambda I\right| = 0 \qquad (4$$

are real since they are the eigenvalues of a symmetric matrix. A characteristic of the system (4.1) is any level surface $\phi(x,t)$ = constant where $\phi(x,t)$ satisfies

$$\det\left|\sum_{j=1}^{n} \phi_{x_j} A_j - \phi_t\right| = 0 \qquad (4.$$

Since the coefficients of (4.7) are Hermitian matrices, the system (4.7) can be written as a symmetric hyperbolic system of 2m real equations in 2n + 1 real independent variables, independent of the type of the system (4.1). Such problems are well posed (c.f. [20] pp. 434-448).

The characteristic surfaces of (4.7) are real manifolds of dimension 2n and define the domain of dependence in the initial hyperplane t = 0 where data must be known if the value of the solution at a given point is to be determined ([20], pp. 614-621).

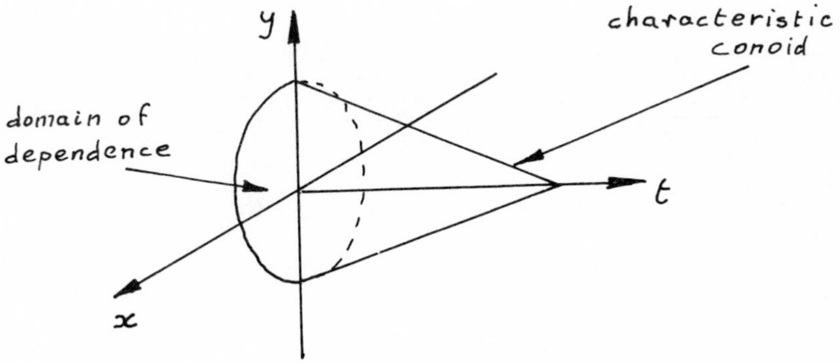

figure 4.1

5. Uniqueness of Solutions to Cauchy's Problem and the Runge Approximation Property

Let L be a second order elliptic operator with analytic coefficients and Laplacian as its principal part. Let M be the adjoint operator.

Definition 5.1: Solutions of an equation Lu = 0 are said to have the Runge approximation property if, whenever D_1 and D_2 are two bounded simply connected domains, D_1 a subset of D_2, any solution in D_1 can be approximated uniformly in compact subsets of D_1 by a sequence of solutions which can be extended as solutions to D_2.

Theorem 5.1 ([30], [32]): Solutions of Lu = 0 have the Runge approximation property if and only if solutions of Mu = 0 are uniquely determined throughout their domain of existence by their Cauchy data along any smooth hypersurface.

Remarks: From a result of Friedrichs ([18]) the L_2 norm of a solution u over a domain bounds the maximum norm of u over any compact subset of this domain. Hence in order to show that a solution over D_1 can be approximated uniformly over any compact subset by solutions in D_2, it is sufficient to show that it can be approximated in the L_2 sense over any subdomain whose closure lies in D_1. Let D_0 be such a subdomai denote by S_1 the restriction of solutions in D_1 to D_0, by S_2 the restriction of solutions in D_2 to D_0. S_2 is a subspace of S_1 (our aim in the first part of the theorem is to show that it is a dense subspace in the L_2 topology). By a classical criterion, S_2 is dense in S_1 if and only if every function v_0 in L_2 over D_0 orthogonal to S_2 is also orthogonal to S_1. Finally we write Green's formula in the form

$$\int_D\int uMw - wLu = \int_C u\frac{\partial w}{\partial \nu} - w\frac{\partial u}{\partial \nu} + cuw \qquad (5.$$

where c is some function of the coefficients of L(see equatio (1.8)).

Proof of Theorem: (uniqueness of solution to Cauchy problem implies Runge approximation property): Let $v_0 \in L_2(D_0)$, $v_0 \perp S_2$. We will show $v_0 \perp S_1$. Let w_0 be a solution of

$$Mw_0 = \begin{cases} v_0 \text{ in } D_0 \\ 0 \text{ in } D_2 - D_0 \end{cases} \qquad (5.$$

$$w_0 = 0 \text{ on } C_2 = \partial D_2.$$

By Theorem D, w_0 exists since $v_0 \perp S_2$. Let $u \in S_2$, $w = w_0$ in (5.1). Since $v_0 \perp S_2$ this shows that $\frac{\partial w_0}{\partial \nu}$ is orthogonal to the boundary values of functions in S_2. By the corollary to Theorem D, $\frac{\partial w_0}{\partial \nu} = \frac{\partial w}{\partial \nu}$, where w satisfies the homogeneous equation $Mw = 0$ in D_2, $w = 0$ on C_2, which implies by Theorem D that (5.2) has a solution w_0 such that $\frac{\partial w_0}{\partial \nu} = 0$ on C_2 (e.g. w -w satisfies this). By the uniqueness of the solution to the

Cauchy problem this implies that $w_0 = 0$ in $D_2 - D_0$.

Now applying (5.1) to $w = w_0$ and $u \in S_1$ over a domain D such that $D_1 \supset D \supset D_0$, we conclude that $v_0 \perp S_1$.

(Runge approximation property implies uniqueness of the solution to the Cauchy problem): Let w_0 be a solution of $Mw_0 = 0$ with zero Cauchy data on a piece of a surface C. We will show $w_0 = 0$ wherever it is defined. First we assume that C is a closed surface and w_0 is defined in a boundary strip of the domain D bounded by C (see figure 5.1).

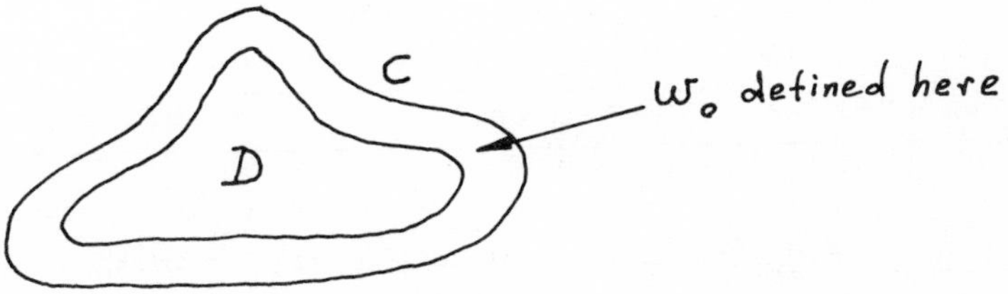

figure 5.1

Since uniqueness is a local problem, without loss of generality choose D so small that L and M are positive definite over D i.e. $Lu = 0$ or $Mu = 0$ in $D_0 \subset D$, $u = 0$ on C, has only the trivial solution for any subdomain D_0.

First we extend w_0 to the whole interior of C (not necessarily as a solution of $Mw_0 = 0$ but such that $Mw_0 \in L_2(D)$). Equation (5.1) with $w = w_0$ and u a solution of $Lu = 0$ shows that $Mw_0 \perp u$ for every such u. By the Runge approximation property this implies that Mw_0 is orthogonal to any solution of $Lu = 0$ in a domain $\tilde{D} \subset D$ which contains the support of $v_0 = Mw_0$. Let $\tilde{C}$ be the (smooth) boundary of $\tilde{D}$.

Now define $\tilde{w}_0$ as the solution of

$$\begin{aligned} M\tilde{w}_0 &= 0 \quad \text{in } \tilde{D} \\ \tilde{w}_0 &= w_0 \quad \text{on } \tilde{C} \end{aligned} \tag{5.3}$$

$\tilde{w}_0$ exists by Theorem D and the fact that L is positive definite. In (5.1) set $w = w_0 - \tilde{w}_0$ and let u be any solution of $Lu = 0$

over $\tilde{D}$ to get

$$\iint\limits_{\tilde{D}} u \, Mw_0 = \int\limits_{\tilde{C}} u \, \frac{\partial(w_0 - \tilde{w}_0)}{\partial \nu} \quad . \qquad (5.4$$

Since $Mw_0 \perp u$, the left-hand side of (5.4) equals zero. Since the boundary values of u on $\tilde{C}$ are arbitrary, we conclude $\frac{\partial(w-\tilde{w}_0)}{\partial \nu} = 0$ on $\tilde{C}$. Now define w_1 by

$$w_1 = \begin{cases} w_0 \text{ in } D - \tilde{D} \\ \tilde{w}_0 \text{ in } \tilde{D} \end{cases} \qquad (5.5$$

Note that w_1 satisfies $Mw_1 = 0$ in both domains and has continuous first derivatives across $\tilde{C}$.

<u>Exercise 5.1</u>. Show that w_1 is a weak solution of $Mw_1 = 0$ and hence a strong (or genuine) solution.

Exercise 5.1 shows that w_1 is an extension of w_0 to the whole interior of C as a solution of $Mw_1 = 0$. Since M is positive definite over D and $w_1 = 0$ on C, $w_1 \equiv 0$ in D which implies $w_0 = 0$ in $D - \tilde{D}$. Since the only restriction on $\tilde{D}$ was that $\tilde{C}$ should be contained in the boundary strip in which w_0 was originally defined, we conclude that $w_0 = 0$ in the whole boundary strip.

We now remove the restriction that C be a closed surface. Let F be a sufficiently small piece of a surface. (F = AB in figure 5.2 below).

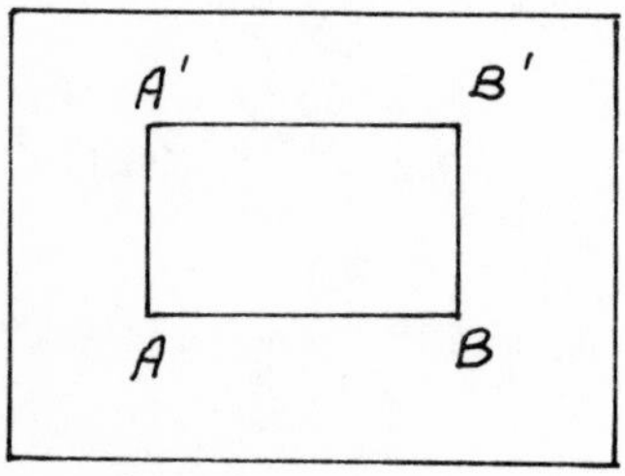

figure 5.2

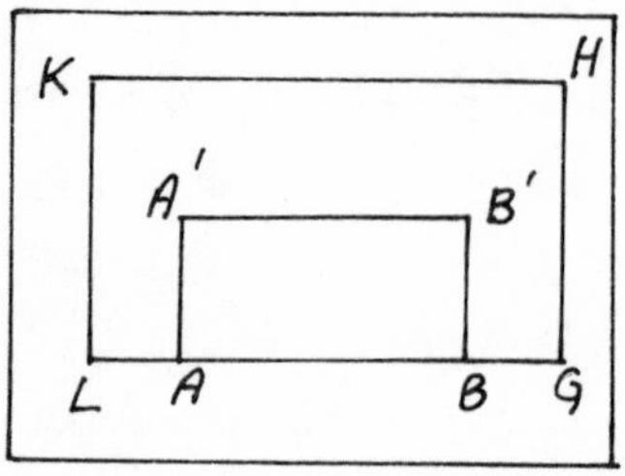

figure 5.3

This yields

$$P^{(2)}(z,z^*) = -2\; Q(z,z^*) \tag{8.9}$$

$$(2n+1)P^{(2n+2)}(z,z^*) = -2\Big(P_z^{(2n)} + Q(z,z^*)\int_0^{z^*} P^{(2n)}(z,z^*)ds^*\Big);$$

$n = 1,2,\ldots$

Thus the $P^{(2n)}$ are uniquely determined. We must now show (8.8) converges.

<u>Definition 8.1</u>: Let $S = \sum_{m,n=0}^{\infty} a_{mn} z^m z^{*n}$, $\tilde{S} = \sum_{m,n=0}^{\infty} \tilde{a}_{mn} z^m z^{*n}$

where $\tilde{a}_{mn} \geq 0$. Then we say the series $\tilde{S}$ dominates the series S if $|a_{mn}| \leq \tilde{a}_{mn}$ $(m,n = 0,1,\ldots)$, and write $S << \tilde{S}$. Note that if $S << \tilde{S}$ then

1) $\frac{\partial S}{\partial z} << \frac{\partial \tilde{S}}{\partial z}$

2) $\int_0^{z^*} S(z,z^*)dz^* << \int_0^{z^*} \tilde{S}(z,z^*)dz^*$

3) $S << \frac{\tilde{S}}{1-az}$, $z \geq 0$.

Since $Q(z,z^*)$ is entire we have $Q(z,z^*) = \sum_{m,n=0}^{\infty} a_{mn} z^m z^{*n}$ converges uniformly and absolutely for $|z| \leq r$, $|z^*| \leq r$ for every $r > 0$. Hence there exists an $M > 0$ such that $|a_{mn} r^m r^n| < M$ $(m,n = 0, 1, 2,\ldots)$, i.e.

$$Q(z,z^*) << M\left(1 - \frac{z}{r}\right)^{-1}\left(1 - \frac{z^*}{r}\right)^{-1} \equiv \tilde{Q}(z,z^*). \tag{8.10}$$

Now define $\tilde{P}^{(2n)}(z,z^*)$ $(n = 1, 2, \ldots)$ by

$$\tilde{P}^{(2)}(z,z^*) = 2\;\tilde{Q}(z,z^*)$$

$$(2n+1)\tilde{P}^{(2n+2)}(z,z^*) = 2\left(\tilde{P}_z^{(2n)}(z,z^*)\left(1 - \frac{z^*}{r}\right)^{-1} + \right.$$

$$\left. + \tilde{Q}(z,z^*)\int_0^{z^*} \tilde{P}^{(2n)}(z,z^*)\left(1 - \frac{z^*}{r}\right)^{-1} dz^* \right. +$$

$$+ c^{(2n)} rMn^{-1} \left(1 - \frac{z}{r}\right)^{-n-1}\left(1 - \frac{z^*}{r}\right)^{-n-1}\Big), \quad (n = 1,2,\ldots) \qquad (8.1$$

where

$$c^{(2)} = 2M$$

$$c^{(2m+2)} = c^{(2n)}\left(\frac{2n}{2n+1}\,\frac{1}{r} + \frac{2Mr}{n(2n+1)}\right), \quad (n=1,2,\ldots) \qquad (8.1$$

Note that the $\tilde{P}^{(2n)}$ are uniquely determined and $P^{(2n)} << \tilde{P}^{(2n}$

Exercise 8.1: Show that

$$\tilde{P}^{(2n)}(z,z^*) = c^{(2n)}\left(1 - \frac{z}{r}\right)^{-n}\left(1 - \frac{z^*}{r}\right)^{-n}, \quad (n = 1,2 \ldots)$$

where for each $\epsilon > 0$

$$c^{(2n)} \leq N\left(\frac{1+\epsilon}{r}\right)^{n}, \quad (n = 1,2,\ldots) \text{ for some } N = N(\epsilon)$$

Exercise 8.1 implies that

$$\tilde{P}^{(2n)} << \frac{N(1+\epsilon)^n}{r^n\left(1 - \frac{z}{r}\right)^n\left(1 - \frac{z^*}{r}\right)^n}, \quad (n = 1, 2, \ldots) \qquad (8.1$$

Hence

$$1 + N\int_0^{z^*} \sum_{n=1}^{\infty} \frac{z^n(1+\epsilon)^n}{r^n\left(1 - \frac{z}{r}\right)^n\left(1 - \frac{z^*}{r}\right)^n}\, dz^* \qquad (8.14$$

is a dominant for (8.8) with $|t| \leq 1$. Since ϵ is arbitrary (8.14) will converge uniformly and absolutely provided that

$$\left|\frac{z}{r\left(1 - \frac{|z|}{r}\right)\left(1 - \frac{|z^*|}{r}\right)}\right| \leq \eta < 1. \qquad (8.15$$

Since Q is entire, r can be arbitrarily large which implies that the series (8.8) converges to an entire function of z and

* for $|t| \leq 1$. Q.E.D.

Since $q(x,y)$ is real valued for x,y real, Re $\underset{\sim}{P}_2\{f\}$ is a olution of $\Delta_2 u - q(x,y)u = 0$. For x, y real we have

$$\text{Re } \underset{\sim}{P}_2\{f\} = \frac{1}{2}\left[\int_{-1}^{1} E(z, \bar{z}, t) f\left(\frac{z}{2}(1 - t^2)\right) \frac{dt}{\sqrt{1-t^2}} + \int_{-1}^{1} \bar{E}(\bar{z}, z, t)\, \bar{f}\left(\frac{\bar{z}}{2}(1-t^2)\right) \frac{dt}{\sqrt{1-t^2}}\right] \tag{8.16}$$

here $\bar{f}(z) = \overline{f(\bar{z})}$, $\bar{E}(z,\bar{z},t) = \overline{E(\bar{z}, z, t)}$. Now extend x and y nto the complex plane, i.e. set $\bar{z} = z^*$ in (8.16). From (8.7) e have

$$\text{Re } \underset{\sim}{P}_2\{f\}\Big|_{z^*=0} = \frac{1}{2}\left[\pi\bar{f}(0) + \int_{-1}^{1} f\left(\frac{z}{2}(1-t^2)\right) \frac{dt}{\sqrt{1-t^2}}\right] \tag{8.17}$$

ow let $u(x,y)$ be a real valued C^2 solution of $\Delta_2 u - q(x,y)u = 0$ n D. Then we can (locally) expand $U(z,z^*) = u\left(\frac{z+z^*}{2}, \frac{z-z^*}{2i}\right)$ as

$$U(z,z^*) = \sum_{m,n=0}^{\infty} a_{mn}\, z^m\, z^{*n} \tag{8.18}$$

ince $U(z,z^*)$ is real valued for x, y real, $a_{mn} = \overline{a_{nm}}$

xercise 8.2: Show from equation (8.3) and the fact that $a_{mn} = \overline{a_{nm}}$ that $U(z,z^*)$ is uniquely determined by $U(z,0)$.

xercise 8.2 shows that if we choose $f(z)$ in equation (8.17) uch that Re $\underset{\sim}{P}_2\{f\}\Big|_{z^*=0} = U(z,0)$, then we have

$$u(x,y) = \text{Re } \underset{\sim}{P}_2\{f\} \tag{8.19}$$

.e. <u>every</u> real valued C^2 solution of $\Delta_2 u - q(x,y)u = 0$ can be xpressed in the form of equation (8.19) for some analytic unction $f(z)$.

heorem 8.3 ([2]): The functions $\{\text{Re } \underset{\sim}{P}_2\{z^n\}\}_{n=0}^{\infty}$ and

$\{\text{Im } P_2\{z^n\}\}_{n=0}^{\infty}$ together form a complete family of solutions for $\Delta_2 u - q(x,y)u = 0$ in any simply connected domain D.

Proof: The proof follows from (8.19) Theorem B, and Runge's Theorem for analytic functions of a single complex variable.

Remark: A similar analysis (c.f. [1]) as in this section yield integral operators for the equation

$$\Delta_2 u + a(x,y)u_x + b(x,y)u_y + c(x,y)u = 0 \qquad (8.20)$$

Alternatively, integral operators for equation (8.20) can be obtained via the Riemann function, c.f. Section 1, exercise 3.1 and equation (11.7). Equation (11.7) defines an operator mapping ordered pairs of analytic functions onto complex valued solutions of equation (8.20). Taking the real parts of both sides of equation (11.7) yields an operator mapping a single analytic function onto real valued solutions of equation (8.20) (c.f. [39]).

9. Integral Operators for Self Adjoint Equations in Three Independent Variables

Consider the partial differential equation

$$\Delta_3 u - q(x,y,z)u = 0 \qquad (9.1)$$

where $q(x,y,z)$ is a real valued (for x, y, z real) entire function of the (complex) variables x, y, z.

Theorem 9.1 ([8]):

Let

$$\begin{aligned} X &= x \\ Z &= \tfrac{1}{2}(y + iz) \\ Z^* &= \tfrac{1}{2}(-y + iz) \end{aligned} \qquad (9.2)$$

nd let $u(x,y,z)$ be a real valued C^2 solution of (9.1) in a eighbourhood of the origin. Then $\mathcal{U}(X,Z,Z^*) = u(x,y,z)$ is an nalytic function of X, Z, Z^* in some neighbourhood of the rigin in C_3 and is uniquely determined by the function $(X,Z^*) = \mathcal{U}(X,0,Z^*)$.

roof: $u(x,y,z) \in C^2$ implies that $\mathcal{U}(X,Z,Z^*)$ is analytic. Hence ocally

$$\mathcal{U}(X,Z,Z^*) = \sum a_{mn\ell} X^\ell Z^n Z^{*m} \tag{9.3}$$

$$\mathcal{U}(X,0,Z^*) = \sum a_{mo\ell} X^\ell Z^{*m} \tag{9.4}$$

$$\mathcal{U}(X,Z,0) = \sum a_{on\ell} X^\ell Z^n.$$

(x,y,z) real valued implies that for x, y, z real

$$\mathcal{U}(X,Z,Z^*) = \overline{\mathcal{U}(X,Z,Z^*)} \tag{9.5}$$

nd hence for x, y, z real

$$\sum a_{mn\ell} X^\ell Z^n Z^{*m} = \sum \overline{a_{mn\ell}} X^\ell (-Z^*)^n (-Z)^m, \tag{9.6}$$

.e.

$$a_{mn\ell} = (-1)^{n+m} \overline{a_{nm\ell}} . \tag{9.7}$$

quations (9.7) and (9.4) imply that $\mathcal{U}(X,Z,0)$ is uniquely etermined from $\mathcal{U}(X,0,Z^*)$. But in X, Z, Z^* coordinates (9.1) ecomes

$$\mathcal{U}_{XX} - \mathcal{U}_{ZZ^*} - Q(X,Z,Z^*) = 0 \tag{9.8}$$

where for x,y, z real $Q(X,Z,Z^*) = q(x,y,z)$). Hence from heorem F (see also section 7) $\mathcal{U}(X,Z,Z^*)$ is uniquely determined rom $\mathcal{U}(X,Z,0)$ and $\mathcal{U}(X,0,Z^*)$ i.e. from $\mathcal{U}(X,0,Z^*)$ alone.

Now define

$$\xi_1 = 2\zeta Z \tag{9.9}$$

$$\xi_2 = X + 2\zeta Z$$

$$\xi_3 = X + 2\zeta^{-1}Z^* \tag{9.9}$$

$$\mu = \tfrac{1}{2}(\xi_2 + \xi_3) = X + \zeta Z + \zeta^{-1}Z^* \tag{9.1}$$

where

$$1 - \epsilon < |\zeta| < 1 + \epsilon, \; 0 < \epsilon < \tfrac{1}{2}.$$

Theorem 9.2 ([8]): Let D be a neighbourhood of the origin in the μ plane, $B = \{\zeta : 1 - \epsilon < |\zeta| < 1 + \epsilon\}$, G a neighbourh of the origin in the ξ_1, ξ_2, ξ_3 space, and $T = \{t : |t| \leq 1\}$. Let $f(\mu,\zeta)$ be an analytic function of two complex variables i $D \times B$ and let $E^*(\xi_1,\xi_2,\xi_3,\zeta,t) \equiv E(X,Z,Z^*,\zeta,t)$ be a regular solution in $G \times B \times T$ of the partial differential equation

$$\mu t(4E^*_{13} + 2E^*_{23} - E^*_{22} - E^*_{33} + Q^*E^*) + (1-t^2)E^*_{1t} - \frac{1}{t}E^*_1 = \tag{9.1}$$

where

$$Q^*(\xi_1,\xi_2,\xi_3,\zeta) \equiv Q(X,Z,Z^*) \text{ and } E^*_i = \frac{\partial E}{\partial \xi_i}.$$

Then

$$U(X,Z,Z^*) = \underset{\sim}{P}_3\{f\} =$$

$$= \frac{1}{2\pi i}\int_{|\zeta|=1}\int_{-1}^{+1} E(X,Z,Z^*,\zeta,t)f(\mu(1-t^2),\zeta)\frac{dt}{\sqrt{1-t^2}}\frac{d\zeta}{\zeta} \tag{9.1}$$

is a (complex valued) solution of (9.1) which is regular in a neighbourhood of the origin in X, Z, Z* space.

Proof: The Jacobian of the transformation (9.9) is -4 which implies that $U(X,Z,Z^*) = \underset{\sim}{P}_3\{f\}$ is regular in a neighbourhood the origin. Differentiating and integrating by parts in (9.12 (using $\frac{\partial f}{\partial w} = \frac{-1}{2\mu t}\frac{\partial f}{\partial t}$ where $w = \mu(1-t^2)$) leads to

$$U_{zz^*} - U_{xx} + QU =$$

$$= \frac{1}{2\pi i} \int_{|\zeta|=1} \int_{-1}^{+1} \frac{f(\mu(1-t^2),\zeta)}{\mu t} \{\mu t(4E^*_{13} + 2E^*_{23} - E^*_{33} + Q^*E^*)$$

$$+ (1-t^2)E^*_{1t} - \frac{1}{t} E^*_1\} \frac{dt}{\sqrt{1-t^2}} \frac{d\zeta}{\zeta}$$

hich implies the theorem.

heorem 9.3([8]): Let $D_r = \{(\xi_1, \xi_2, \xi_3) + |\xi_i| < r, i = 1,2,3\}$ here r is an arbitrary positive number, and $B_{2\epsilon}=\{\zeta : |\zeta-\zeta_0|<2\epsilon\}$, $< \epsilon < \frac{1}{2}$, where ζ_0 is arbitrary with $|\zeta_0| = 1$. Then for every , n = 0, 1, 2, ... there exists a unique function $^{(n)}(\xi_1, \xi_2, \xi_3, \zeta)$ regular in $\overline{D}_r \times \overline{B}_{2\epsilon}$ and satisfying

$$p_1^{(n+1)} = \frac{1}{2n+1}\left[p_{22}^{(n)} - p_{33}^{(n)} - 4p_{13}^{(n)} - 2p_{23}^{(n)} - Q^*p^{(n)}\right]$$

$$p^{(0)}(\xi_1, \xi_2, \xi_3, \zeta) \equiv 1, \quad p^{(n+1)}(0, \xi_2, \xi_3, \zeta) = 0;$$

$$n = 0,1,2,... \tag{9.13}$$

here $p_i^{(n)} = \frac{\partial p^{(n)}}{\partial \xi_i}$. Furthermore the function

$$E^*(\xi_1, \xi_2, \xi_3, \zeta,t) = 1 + \sum_{n=1}^{\infty} t^{2n}\mu^n p^{(n)}(\xi_1,\xi_2,\xi_3,\zeta) \tag{9.14}$$

s a solution of (9.11) which is regular in $G_R \times B \times T$ where R s an arbitrary positive number and

$$G_R = \{(\xi_1,\xi_2,\xi_3) + |\xi_i| < R, i = 1,2,3\}$$

$$B = \{\zeta : 1 - \epsilon < |\zeta| < 1 + \epsilon\}, \ 0 < \epsilon < \tfrac{1}{2}$$

$$T = \{t : |t| \leq 1\}.$$

he function defined in (9.14) satisfies

$$E^* (0, \xi_2 \ \xi_3 \ \zeta, t) = 1 \tag{9.15}$$

<u>Proof</u>: $p^{(1)}(\xi_1,\xi_2,\xi_3,\zeta) = -\int_0^{\xi} Q^*(\xi_1',\xi_2\ \xi_3\ \zeta)d\xi_1'$ is uniquely determined and is regular in $\overline{D}_r \times \overline{B}_{2\epsilon}$. By induction all the $p^{(n)}$ are uniquely determined. Substituting (9.14) into (9.11) shows that E* formally satisfies (9.11). We must now show the series converges uniformly in $G_R \times B \times T$. Since $\overline{B}$ is compact there exists $\zeta_j, |\zeta_j| = 1$, $j = 1, \ldots, N$ such that B is covered by $\bigcup_{j=1}^{N} N_j$ where $N_j = \{\zeta - \zeta_j| < \frac{3}{2}\epsilon\}$. Hence it is sufficient to show that the series converges in $\overline{G}_R \times \overline{N}_j \times T$. Since Q is entire, in $\overline{D}_r \times \overline{B}_{2\epsilon}$ we have

$$Q^*(\xi_1,\xi_2,\xi_3,\zeta) << C\left(1-\frac{\xi_1}{r}\right)^{-1}\left(1-\frac{\xi_2}{r}\right)^{-1}\left(1-\frac{\xi_3}{r}\right)^{-1}\left(1-\frac{\zeta-\zeta_0}{2\epsilon}\right)^{-1}$$

for some C > 0 where "<<" means "is dominated by".

<u>Exercise 9.1</u>: Show by induction that in $\overline{D}_r \times \overline{B}_{2\epsilon}$

$$p_1^{(n)} << M(8+\delta)^n(2n-1)^{-1}\left(1-\frac{\xi_1}{r}\right)^{-(2n-1)}\left(1-\frac{\xi_2}{r}\right)^{-(2n-1)}$$
$$\left(1-\frac{\xi_3}{r}\right)^{-(2n-1)}\left(1-\frac{\zeta-\zeta_0}{2\epsilon}\right)^{n} r^{-n}$$

where M and δ are positive constants independent of n.

Exercise 9.1 implies that

$$p^{(n)} << M(8+\delta)^n(2n)^{-1}(2n-1)^{-1}\left(1-\frac{\xi_1}{r}\right)^{-2n}\left(1-\frac{\xi_2}{r}\right)^{-(2n-1)}$$
$$\left(1-\frac{\xi_3}{r}\right)^{-(2n-1)}\left(1-\frac{\zeta-\zeta_0}{2\epsilon}\right)^{-n} r^{-n+1}$$

and hence in $\overline{D}_r \times \overline{N}_j \times T$ we have

$$|p^{(n)}(\xi_1,\xi_2,\xi_3,\zeta)| \le M(8+\delta)^n(2n)^{-1}(2n-1)^{-2}\left(1-\frac{|\xi_1|}{r}\right)^{-2n}$$
$$\left(1-\frac{|\xi_2|}{r}\right)^{-(2n-1)}\left(1-\frac{|\xi_3|}{r}\right)^{-(2n-1)}\left(1-\frac{|\zeta-\zeta_j|}{2\epsilon}\right)^{-n} r^{-n+1}.$$

Now consider $|t^{2n} \mu^n p^{(n)}(\xi_1,\xi_2,\xi_3,\zeta)|$ in $\overline{D}_{\alpha r} \times \overline{N}_j \times T$ where

$$D_{\alpha r} = \{(\xi_1,\xi_2,\xi_3) : |\xi_i| < \frac{r}{\alpha} ; \alpha > 1, i = 1,2,3\}$$

In $\overline{D}_{\alpha r} \times \overline{N}_j \times T$ we have

$$1 - \frac{|\xi_i|}{r} \geq \frac{\alpha-1}{\alpha} , \quad i = 1, 2, 3$$

$$1 - \frac{|\zeta-\zeta_j|}{2\epsilon} \geq \frac{1}{4}$$

$$|\mu| = \tfrac{1}{2}|\xi_2 + \xi_3| \leq \frac{r}{\alpha}$$

$$|t| \leq 1$$

which implies

$$|t^{2n}\mu^n p^{(n)}(\xi_1,\xi_2,\xi_3,\zeta)| \leq Mr(\alpha-1)^2\alpha^{-2}(2n)^{-1}(2n-1)^{-1} \{4 \alpha^5(\delta+\delta)(\alpha-1)^{-6}\}^n$$

and hence for α sufficiently large the series (9.14) converges absolutely and uniformly in $\overline{D}_{\alpha r} \times \overline{N}_j \times T$. Setting $r = \alpha R$ shows that E^* is regular in $\overline{G}_R \times \overline{N}_j \times T$ and hence is regular in $G_R \times B \times T$.

Exercise 9.2: Show that if $Q(X,Z,Z^*) = \lambda$ = constant, then

$$E(X,Z,Z^*,\zeta,t) = \cos \sqrt{4\lambda(\zeta XZ + \zeta^2 Z^2 + ZZ^*)}\, t.$$

Exercise 9.3: Show that $E(X,Z,Z^*,\zeta,t)$ is an entire function of its five independent complex variables if $Q(X,Z,Z^*)$ is entire.

Theorem 9.4 ([8]): Let $u(x,y,z)$ ($= U(X,Z,Z^*)$) be a real valued C^2 solution of (9.1) in some neighbourhood of the origin in $\mathbb{R}^3$. Then there exists an analytic function of two complex variables $f(\mu,\zeta)$ which is regular for $|\zeta| < 1 + \epsilon, \epsilon > 0$, and μ in some neighbourhood of the origin such that locally $u(x,y,z) = \text{Re}\, \underset{\sim}{P}_3\{f\}$.

In particular

$$f(\mu,\zeta) = -\frac{1}{2\pi}\int_\gamma g(\mu(1-t^2),\zeta)\,\frac{dt}{t^2}$$

where

$$g(\mu,\zeta) = 2\,\frac{\partial}{\partial\mu}\left[\mu\int_0^1 \mathbf{U}(t\mu,\ 0,\ (1-t)\mu\zeta)dt\right] - \mathbf{U}(\mu,0,0) \qquad (9.1$$

and γ is a rectifiable arc joining the points $t = -1$ and $t = +$ and not passing through the origin.

<u>Proof</u>: $u(x,y,z)\in C^2$ implies that $u(x,y,z)$ is analytic and $q(x,y,z)$ real implies that Re $\underset{\sim}{P}_3\{f\}$ (where x,y,z are real) is a real valued solution of (9.1) for every analytic function $f(\mu,\zeta)$. Now suppose locally that

$$g(\mu,\zeta) = \sum_{n=0}^{\infty}\sum_{m=0}^{n} a_{nm}\,\mu^n\,\zeta^m$$

$$f(\mu,\zeta) = -\frac{1}{2\pi}\int_\gamma g(\mu(1-t^2),\zeta)\,\frac{dt}{t^2} \qquad (9.1$$

$$= \sum_{n=0}^{\infty}\sum_{m=0}^{n} a_{nm}\,\frac{\Gamma(n+1)}{\Gamma(n+\frac{1}{2})\Gamma(\frac{1}{2})}\,\mu^n\zeta^m$$

$$Q(X,Z,Z^*) = \sum_{\ell,m,n=0}^{\infty} b_{mn}\,X^\ell Z^n Z^{*m}.$$

<u>Exercise 9.4</u>: Show that

$$g(\mu,\zeta) = \int_{-1}^{1} f(\mu(1-t^2),\zeta)\,\frac{dt}{\sqrt{1-t^2}}\ .$$

Define

$$\overline{Q}(X,Z,Z^*) = \sum_{\ell,m,n=0}^{\infty} \overline{b_{mn\ell}}\,X^\ell Z^n Z^{*m}$$

$$\overline{f}(\mu,\zeta) = \sum_{n=0}^{\infty}\sum_{m=0}^{n} \overline{a_{nm}}\,\frac{\Gamma(n+1)}{\Gamma(n+\frac{1}{2})\Gamma(\frac{1}{2})}\,\mu^n\,\zeta^m. \qquad (9.1$$

et $\overline{E}$ (X,Z,Z^*,ζ,t) be the generating function corresponding o the equation $U_{xx} - U_{zz^*} - \overline{Q}U = 0$. Then for x,y,z real

$$\mathrm{Re}\ \underset{\sim}{P}_3\{f\} = \frac{1}{4\pi i}\int_{|\zeta|=1}\int_{-1}^{1} E(X,Z,Z^*,\zeta,t)f(\mu(1-t^2),\zeta)\frac{dt}{\sqrt{1-t^2}}\frac{d\zeta}{\zeta}$$

$$+\frac{1}{4\pi i}\int_{|\zeta|=1}\int_{-1}^{1} \overline{E}(X,-Z^*,-Z,\zeta,t)\overline{f}(\overline{\mu}(1-t^2),\zeta)\frac{dt}{\sqrt{1-t^2}}\frac{d\zeta}{\zeta} \tag{9.19}$$

where $\overline{\mu} = X - \zeta Z^* - \zeta^{-1}Z$. From Theorem 9.1 (X,Z,Z^*) is uniquely determined by $U(X,0,Z^*)$ and hence we try and determine $f(\mu,\zeta)$ from the integral equation

$$U(X,0,Z^*) = \frac{1}{4\pi i}\int_{|\zeta|=1}\int_{-1}^{1} f(\mu_1(1-t^2),\zeta)\frac{dt}{\sqrt{1-t^2}}\frac{d\zeta}{\zeta}$$

$$+\frac{1}{4\pi i}\int_{|\zeta|=1}\int_{-1}^{1}\overline{E}(X-Z^*,0,\zeta,t)\overline{f}(\mu_2(1-t^2),\zeta)\frac{dt}{\sqrt{1-t^2}}\frac{d\zeta}{\zeta} \tag{9.20}$$

where $\mu_1 = X + \zeta^{-1}Z^*$, $\mu_2 = X - \zeta Z^*$. But

$$\overline{E}^*(\xi_1,\xi_2,\xi_3,\zeta,t) = 1 + \sum_{n=1}^{\infty} t^{2n}\mu^n\,\overline{p}^{(n)}(\xi_1,\xi_2,\xi_3,\zeta) \tag{9.21}$$

where $\overline{p}^{(1)} = -\int_0^{\xi_1}\overline{Q}^*(\xi_1',\xi_2,\xi_3,\zeta)d\xi_1'$

$$\overline{p}_1^{(n+1)} = \frac{1}{2n+1}\{\overline{p}_{22}^{(n)} + \overline{p}_{33}^{(n)} - 4\overline{p}_{13}^{(n)} - 2\overline{p}_{23}^{(n)} - \overline{Q}^*\overline{p}^{(n)}\}$$

$$\overline{p}^{(n+1)}(0,\xi_2,\xi_3,\eta) = 0;\ n = 0,1,2,\ldots \tag{9.22}$$

Equations (9.9), (9.22) imply that

$$\overline{p}^{(1)} = -2\zeta\int_0^{Z}\overline{Q}(X + 2\zeta Z - 2\zeta\tau,\ \tau,\ \tfrac{\zeta}{2}(2\zeta^{-1}Z^* - 2\zeta Z + 2\zeta\tau))d\tau$$

i.e. $\overline{p}^{(1)}$ is an entire function of X, Z, Z* and ζ, and vanish for $\zeta = 0$. A similar calculation using (9.22) shows that the same can be said for $\overline{p}^{(n)}$ $n = 1, 2, \ldots$.

Substituting (9.21) into (9.20) and integrating termwise (this is possible due to the absolute and uniform convergence of the series (9.21)) gives

$$\begin{aligned} U(X,0,Z^*) &= \frac{1}{4\pi i}\int_{|\zeta|=1}\int_{-1}^{1} f(\mu_1(1-t^2),\zeta)\,\frac{dt}{\sqrt{1-t^2}}\,\frac{d\zeta}{\zeta} \\ &+ \frac{1}{4\pi i}\int_{|\zeta|=1}\int_{-1}^{1} \overline{f}(\mu_2(1-t^2),\zeta)\,\frac{dt}{\sqrt{1-t^2}}\,\frac{d\zeta}{\zeta} \\ &= \frac{1}{4\pi i}\int_{|\zeta|=1} g(\mu_1,\zeta)\,\frac{d\zeta}{\zeta} \\ &+ \frac{1}{4\pi i}\int_{|\zeta|=1} \overline{g}(\mu_2,\zeta)\,\frac{d\zeta}{\zeta} \end{aligned} \tag{9.23}$$

where

$$\overline{g}(\mu,\zeta) = \sum_{n=0}^{\infty}\sum_{m-0}^{n} \overline{a_{nm}}\;\mu^n\,\zeta^m.$$

We will now show that (9.16) gives the solution of (9.23). Let

$$U(X,0,Z^*) = \sum_{n=0}^{\infty}\sum_{m=0}^{\infty} c_{nm}\, X^n Z^{*m} \tag{9.24}$$

Since $u(x,y,z)$ is real valued we have that c_{no}, $n = 0,1,2,\ldots$ are all real. Equating coefficients of $X^n\,Z^{*m}$ in (9.23) gives

$$\begin{aligned} 2n!m!\,c_{nm} &= (n+m)!\,a_{n+m,m}\ ;\ n \geq 0,\ m > 0 \\ 2c_{no} &= a_{no} + \overline{a}_{no} \end{aligned} \tag{9.25}$$

Without loss of generality assume that a_{no}, $n = 0,1,2,\ldots$

e real. Equations (9.24), (9.25) imply that

$$\mathcal{U}(X,0,Z^*) = \frac{1}{2}\sum_{m=0}^{\infty}\sum_{n=0}^{\infty}\frac{\Gamma(n+m+1)}{\Gamma(n+1)\Gamma(m+1)}\, a_{n+m,m}X^nZ^{*m} + \frac{1}{2}\sum_{n=0} c_{no}X^n$$

$$= \frac{1}{2}\sum_{m=0}^{\infty}\sum_{n=m}^{\infty}\frac{\Gamma(n+1)}{\Gamma(n-m+1)\Gamma(m+1)}\, a_{nm}X^{n-m}Z^{*m} + \frac{1}{2}\sum_{n=0}^{\infty} c_{no}\, X^n \tag{9.26}$$

$$= \frac{1}{2}\sum_{n=0}^{\infty}\sum_{m=0}^{n}\frac{\Gamma(n+1)}{\Gamma(n-m+1)\Gamma(m+1)}\, a_{nm}\, X^{n-m}Z^{*m} + \frac{1}{2}\sum_{n=0}^{\infty} c_{no}X^n.$$

om the definition of the Beta function B(x,y)

$$B(x,y) \equiv \int_0^1 t^{x-1}(1-t)^{y-1}\, dt = \frac{\Gamma(x)\Gamma(y)}{\Gamma(x+y)}$$

have

$$\int_0^1 \mathcal{U}(tX,0,(1-t)Z^*)dt = \frac{1}{2}\sum_{n=0}^{\infty}\sum_{m=0}^{n}\frac{a_{nm}}{n+1}\, X^{n-m}Z^{*m} + \frac{1}{2}\sum_{n=0}^{\infty}\frac{c_{no}}{n+1}\, X^n$$

$$\frac{\partial}{\partial\mu}\left[\mu\int_0^1 \mathcal{U}(t\mu,0,(1-t)\mu\zeta)dt\right] =$$

$$= \frac{1}{2}\sum_{n=0}^{\infty}\sum_{m=0}^{n} a_{nm}\, \mu^n\zeta^m + \frac{1}{2}\sum_{n=0}^{\infty} c_{no}\mu^n$$

$$= \frac{1}{2}g(\mu,\zeta) + \frac{1}{2}\,\mathcal{U}(\mu,0,0)$$

which implies the theorem.

<u>Theorem 9.5 ([8])</u>: Let G be a bounded, simply connected domai in $\mathbb{R}^3$, and define

$$u_{2n,m}(x,y,z) = \text{Re } \underset{\sim}{P}_3\{\mu^n\zeta^m\};\ 0 \leq n < \infty,\ m=0,\ 1,\dots\ n.$$
(9.27
$$u_{2n+1,m}(x,y,z) = \text{Im } \underset{\sim}{\mathbf{P}}_3\{\mu^n\zeta^m\};\ 0 \leq n < \infty,\ m=0,\ 1,\dots,n.$$

Then the set $\{u_{nm}\}$ is a complete family of solutions for equation (9.1) in the space of real valued C^2 solutions of (9.1) defined in G.

<u>Proof</u>: Let $u(x,y,z)\in C^2$ be a solution of (9.1) in G and let $\overline{G}_1 \subset G$. By the Runge approximation property for every $\epsilon > 0$ there exists a solution $u_1(x,y,z)$ of (9.1) which is regular in a sphere S, $S \supset G$, such that

$$\max_{(x,y,z)\in\overline{G}_1} |u - u_1| < \frac{\epsilon}{3}\ . \qquad (9.28$$

From section 4 we can conclude that the Cauchy data for u_1 mus be regular in some <u>convex</u> region B in C^2 and u_1 depends continuously on this data in S. Since convex domains are Runge domains of the first kind ([19] p.229), on compact subsets of B we can approximate the Cauchy data for u_1 by polynomials and construct a (real valued) solution u_2 of (9.1) with polynomial Cauchy data. By Theorem F $u_2(x,y,z)$ is an entire function of its independent (complex) variables. Furthermore there exists a domain G_2, $G \subset \overline{G}_2 \subset S$, such that

$$\max_{(x,y,z)\in\overline{G}_2} |u_1 - u_2| < \frac{\epsilon}{3}\ . \qquad (9.29$$

The fact that u_2 is entire implies that $\mathbf{U}_2(X,Z,Z^*)$ $(=u_2(x,y,z)$ for x,y,z real) is an entire function of X, Z,Z*, which implie that $\mathbf{U}_2(X,0,Z^*)$ is regular in $\{|X| \leq R\} \times \{|Z^*| \leq R\}$ for R arbitrarily large. Since product domains are Runge domains of the first kind ([19], p.49), we can approximate $\mathbf{U}_2(X,0,Z^*)$

by a polynomial in $\{|x| \leq R\} \times \{|Z^*| \leq R\}$ and use Theorems 9.1 and F to construct a (real valued) entire solution $u_3(x,y,z)$ of equation (9.1) (with polynomial Goursat data in the X, Z, Z* variables) such that

$$\max_{(x,y,z)\in\overline{G}_2} |u_2 - u_3| < \frac{\epsilon}{3}. \tag{9.30}$$

Theorem 9.4 implies that there exists a polynomial $h_N(\mu,\zeta)$ such that $u_3(x,y,z) = \operatorname{Re} \underset{\sim}{P}_3 \{h_N\}$. Equations (9.28) - (9.30) now imply that

$$\max_{(x,y,z)\in\overline{G}_1} |u - \operatorname{Re} \underset{\sim}{P}_3 \{h_N\}| < \epsilon$$

and the theorem follows.

<u>Example 9.1</u>: When $q \equiv 0$ we have

$$u_{2n,m}(x,y,z) = \frac{n!}{(n+m)!} r^n P_n^m (\cos\theta)\operatorname{Re}(i^m e^{im\phi})$$

$$u_{2n+1,m}(x,y,z) = \frac{n!}{(n+m)!} r^n P_n^m (\cos\theta)\operatorname{Im}(i^m e^{im\phi})$$

where r, θ, ϕ are spherical coordinates and P_n^m denote the associated Legendre polynomials.

<u>Remarks</u>: The results in this section first appeared in [8] and [9]. Prior to this paper partial results in this direction had been obtained by Bergman [1], Tjong [38] and Gilbert and Lo [23]. For the extension of the results in this section to elliptic equations in four independent variables see [10]. For recent results in this area see the book by Gilbert referred to in the introduction and the Indiana University Ph.D. thesis of D. Kukral and M. Stecher.

10. <u>Integral Operators for Non Self Adjoint Equations in Three Independent Variables</u>.

The result just presented for equation (9.1) can also be obtained for the more general equation

$$\Delta_3 u + a(x,y,z)u_x + b(x,y,z)u_y + c(x,y,z)u_z + d(x,y,z)u = 0 \tag{10.1}$$

where a, b, c, d are real valued entire functions of the (complex variables) x, y, z(see [9]). In complex form (10.1) becomes

$$\mathcal{U}_{XX} - \mathcal{U}_{ZZ^*} + A(X,Z,Z^*)\mathcal{U}_X + B(X,Z,Z^*)\mathcal{U}_Z + C(X,Z,Z^*)\mathcal{U}_{Z^*} + D(X,Z,Z^*) = 0 \tag{10.2}$$

where $A = a$, $B = \frac{1}{2}(b + ic)$, $C = \frac{1}{2}(-b + ic)$, $D = d$. Substituti

$$V(X,Z,Z^*) = \quad (X,Z,Z^*)e^{-\int_0^Z C(X,Z',Z^*)dZ'} \tag{10.3}$$

yields the following equation satisfied by $V(X,Z,Z^*)$:

$$V_{XX} - V_{ZZ^*} + \tilde{A}(X,Z,Z^*)V_X + \tilde{B}(X,Z,Z^*)V_Z + \tilde{D}(X,Z,Z^*)V = 0 \tag{10.4}$$

where $\tilde{A}$, $\tilde{B}$, $\tilde{D}$ can be expressed in terms of A, B, C. An integral operator mapping analytic functions $f(\mu,\zeta)$ onto solutions $V(X,Z,Z^*)$ of (10.4) is given by

$$\begin{aligned} V(X,Z,Z^*) &\equiv \underset{\sim}{C}_3^! \{f\} \\ &= \frac{1}{2\pi i} \int_{|\zeta|=1} \int_{-1}^{1} E(X,Z,Z^*,\zeta,t) f(\mu(1-t^2),\zeta) \frac{dt}{\sqrt{1-t^2}} \frac{d\zeta}{\zeta} \end{aligned} \tag{10.5}$$

where in the ξ_1,ξ_2,ξ_3 variables $E^* = E$ satisfies

$$\begin{aligned} &\mu t(4E^*_{13} + 2E^*_{23} - E^*_{22} - E^*_{33} - \tilde{D}^*E^*) + (1-t^2)E^*_{1t} \\ &- \frac{1}{t} E^*_1 - \tilde{A}^* [(E^*_2 + E^*_3)\mu t + \frac{1}{2}(1-t^2)E^*_t - \frac{1}{2t} E^*] \\ &- \tilde{B}^*\zeta [(2E^*_1 + 2E^*_2)\mu t + \frac{1}{2}(1-t^2)E^*_t - \frac{1}{2t} E^*] = 0 \end{aligned} \tag{10.6}$$

where $\tilde{A}^*$, $\tilde{B}^*$, $\tilde{D}^*$ are $\tilde{A}$, $\tilde{B}$, $\tilde{D}$ in the ξ_i, i = 1, 2, 3, variables t can be shown as in Theorem 9.3 that

$$E^*(\xi_1,\xi_2,\xi_3,\zeta,t) = \sum_{n=1}^{\infty} t^{2n} \mu^n p^{(n)}(\xi_1,\xi_2,\xi_3,\zeta) \qquad (10.7)$$

s a regular solution of equation (10.6) in $G_R \times B \times T$, where he $p^{(n)}$ are given recursively by

$$p_1^{(n+1)} - \frac{1}{2}(\tilde{A}^* + \tilde{B}^*\zeta)p^{(n+1)} = \frac{1}{2n+1}\{p_{22}^{(n)} - p_{33}^{(n)} - 4p_{13}^{(n)}$$

$$- 2p_{23}^{(n)} + (\tilde{A}^* + 2\tilde{B}^*\zeta)p_2^{(n)} + \tilde{A}^* p_3^{(n)} + 2\tilde{B}^*\zeta\, p_1^{(n)} + \tilde{D}^* p^{(n)}\},$$

$$p^{(1)}(\xi_1,\xi_2,\xi_3,\zeta) = e^{\frac{1}{2}\int_0^{\xi_1}(\tilde{A}^* + \tilde{B}^*\zeta)d\xi'},$$

$$p^{(n+1)}(0,\xi_2,\xi_3,\zeta) = 0; \; n = 1, 2, \ldots \qquad (10.8)$$

In the present case it is not possible to have $E^*(0,\xi_2,\xi_3,\zeta,t)=1$ as in Theorem 9.3 since in this case equation (10.6) cannot be satisfied due to the appearance of the term $\frac{1}{2t}E^*(\tilde{A}^* + \tilde{B}^*\zeta))$.

Proceeding now as in Theorem 9.4 we can show that every real valued solution of equation (10.3) can be represented locally in the form

$$\mathbf{U}(X,Z,Z^*) = \mathbf{U}(0,0,0)\mathbf{U}_0(X,Z,Z^*) + \text{Re}\; \underset{\sim}{C}_3\{f\} \qquad (10.9)$$

where

$$\underset{\sim}{C}_3\{f\} = \frac{1}{2\pi i}\int_{|\zeta|=1}\int_{-1}^{+1} e^{\int_0^Z C(X,Z',Z^*)dZ'} E(X,Z,Z^*,\zeta,t) \times$$

$$f(\mu(1-t^2),\zeta)\frac{dt}{\sqrt{1-t^2}}\frac{d\zeta}{\zeta}, \qquad (10.10)$$

$\mathbf{U}_0(X,Z,Z^*)$ is the unique solution of equation (10.2) satisfying $\mathbf{U}_0(X,0,Z^*) = \mathbf{U}_0(X,Z,0) = 1$ (which can be constructed by iteration - see Theorem F and [28]) and $f(\mu,\zeta)$ is constructed from the formulas

$$f(\mu,\zeta) = \frac{3}{2\pi}\int_{\gamma} g(\mu(1-t^2),\zeta)\,\frac{(1-t^2)}{t^4}\,dt$$

$$g(\mu,\zeta) = \sum_{n=0}^{\infty}\sum_{m=0}^{n+1} a_{nm}\,\mu^n\zeta^m$$

$$a_{n+m-1,m} = \frac{2n!m!}{(n+m)!}\,\gamma_{nm} - \sum_{k=0}^{n-1}\frac{n!}{(n+m)!k!}\,\delta_{km}\gamma_{n-k},0 \qquad (10.11$$

where

$$\mathbf{U}(X,0,Z^*) - \mathbf{U}(0,0,0) = \sum_{n=0}^{\infty}\sum_{\substack{m=0\\ n+m\neq 0}}^{\infty}\gamma_{nm}X^nZ^{*m}$$

$$\delta_{km} = \left(\frac{\partial^{k+m}}{\partial X^k\partial Z^{*m}}\, e^{\int_0^{-Z^*}\overline{C}(X,Z',0)dZ'}\right)_{X=Z^*=0}$$

Remark: It was in order to achieve this inversion formula that equation (10.2) was reduced to the form of equation (10..

Following the analysis of Theorem 9.5 it can now be shown ([9]) that the set

$$u_0(x,y,z) = \mathbf{U}_0(X,Z,Z^*)$$

$$u_{2n,m} = \operatorname{Re} \underset{\sim}{C}_3\,\{\mu^n\zeta^m\};\ 0 \le n < \infty,\ m = 0,1,\ \ldots,\ n+1$$

$$(10.1$$

$$u_{2n+1,m} = \operatorname{Im} \underset{\sim}{C}_3\{\mu^n\zeta^m\};\ 0 \le n < \infty,\ m = 0,1,\ \ldots,\ n+1$$

is a complete family of solutions for equation (10.1).

IV Analytic continuation

11. Lewy's Reflection Principle and Vekua's Integral Operators.

Let u satisfy the partial differential equation

$$u_{xx} + u_{yy} + a(x,y)u_x + b(x,y)u_y + c(x,y)u = 0 \qquad (11.1)$$

where $u(x,y) \in C^2(D) \cap C^1(\bar{D})$, D being a simply connected domain of the x,y plane whose boundary contains a segment σ of the x-axis with the origin as the interior point and such that D contains the portion $y < 0$ of a neighbourhood of each point of σ.

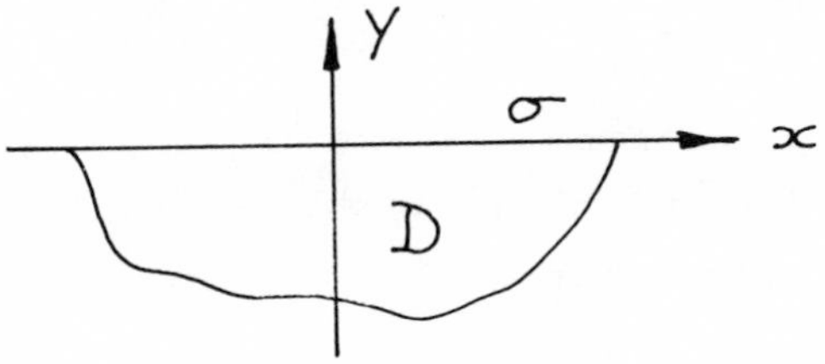

figure 11.1

In complex form (11.1) becomes (c.f. section 1)

$$L[U] = \frac{\partial^2 U}{\partial z \partial z^*} + A(z,z^*)\frac{\partial U}{\partial z} + B(z,z^*)\frac{\partial U}{\partial z^*} + C(z,z^*)U = 0. \qquad (11.2)$$

Assume A, B, C are regular for z, z* in $D \cup \sigma \cup D^* \times D \cup \sigma \cup D^*$. Now let $R(z,z^*,\zeta,\zeta^*)$ be the Riemann function of the adjoint equation (see Section 1 and Exercise 3.1)

$$M[v] = \frac{\partial^2 v}{\partial z \partial z^*} - \frac{\partial (Av)}{\partial z} - \frac{\partial (Bv)}{\partial z^*} + Cv = 0. \qquad (11.3)$$

We have the identity for analytic functions $U(z,z^*)$, $v(z,z^*)$,

$$vL[\mathcal{U}] - \mathcal{U}M[v] = (v\mathcal{U})_{zz^*} - (\mathcal{U}[v_{z^*} - Av])_z - - (\mathcal{U}[v_z - Bv])_{z^*}. \quad (11.$$

Setting $v = R$, letting $\mathcal{U}$ be a solution of $L[\mathcal{U}] = 0$ and usi Green's theorem over a plane region S in C^2 bounded by a smoo curve ∂S gives

$$0 = \oint_{\partial S} (\mathcal{U}R)_z dz - \oint_{\partial S} (R_z - BR)dz + \oint_{\partial S} \mathcal{U}(R_{z^*} - AR)dz^* \quad (11.$$

Letting S be the triangle with corners $(\zeta,\overline{\zeta})$, (ζ,ζ^*) and $(\overline{\zeta}^*,$ gives (setting, at the end, $\zeta = z$, $\zeta^* = z^*$)

$$\begin{aligned}\mathcal{U}(z,z^*) &= \mathcal{U}(\overline{z}^*,z^*)R(\overline{z}^*,z^*,z,z^*) \\ &+ \int_d \mathcal{U}(t,\overline{t})(R_{\overline{t}}(t,\overline{t},z,z^*) - A(t,\overline{t})R)d\overline{t} \\ &+ \int_d [(\mathcal{U}(t,\overline{t})R(t,\overline{t},z,z^*))_t - \mathcal{U}(R_t - BR)]dt\end{aligned} \quad (11.$$

where d is the diagonal from $(\overline{z}^*,z^*)$ to $(z,\overline{z})$ and use has bee made of the boundary conditions satisfied by R. Similarly, letting S be the quadrilateral $(0,0)$, $(\zeta,0)$, $(\zeta,\overline{\zeta})$, $(0,\overline{\zeta})$, we have

$$\begin{aligned}\mathcal{U}(z,\overline{z}) &= -\mathcal{U}(0,0)R(0,0,z,\overline{z}) + \mathcal{U}(z,0)R(z,0,z,\overline{z}) \\ &+ \mathcal{U}(0,\overline{z})R(0,\overline{z},z,\overline{z}) \\ &- \int_0^z \mathcal{U}(t,0)(R_t(t,0,z,\overline{z}) - B(t,0)R)dt \\ &- \int_0^{\overline{z}} \mathcal{U}(0,t)(R_{\overline{t}}(0,\overline{t},z,\overline{z}) - A(0,\overline{t})R)d\overline{t}.\end{aligned} \quad (11.$$

(Equation (11.7) is Vekua's operator ([39]) mapping analytic functions onto solutions of equation (11.1)).

Now suppose on σ we have

$$u(x,0) = \mathcal{U}(x,x) = \delta(x) \quad (11.$$

here $\delta(z)$ is regular for z in $D \cup \sigma \cup D^*$. In equation (11.7) et $\mathcal{U}(z,0) = f(z)$, $\mathcal{U}(0,z) = g(z)$, and obtain for z on σ

$$\begin{aligned}\delta(z) &= -\delta(0)R(0,0,z,z) + f(z)R(z,0,z,z) + g(z)R(0,z,z,z) \\ &- \int_0^z f(t)(R_t(t,0,z,z) - B(t,0)R)dt \\ &- \int_0^z g(t)(R_t(0,t,z,z) - B(0,t)R)dt\end{aligned} \tag{11.9}$$

ote that the boundary conditions satisfied by R imply that $(z,0,z,z)$ and $R(0,z,z,z)$ do not vanish for z in $D \cup \sigma \cup D^*$. rom equation (11.6) (setting $z^* = 0$) we see that $f(z)$ is known or z in $D \cup \sigma$ in terms of the given solution $u(x,y) = \mathcal{U}(z,\bar{z})$ nd(setting $z = 0$) so in $g(z)$ for z in $D^* \cup \sigma$.

Now for z in $D \cup \sigma$ (11.9) is a Volterra integral equation or $g(z)$ since $f(z)$ and $\delta(z)$ are known in $D \cup \sigma$. Since the ernel and terms not involving $g(z)$ are analytic in D, con- inuous in $D \cup \sigma$, so must the solution $g(z)$. But $g(z)$ is already nown to be analytic in D^* and continuous in $D^* \cup \sigma$, which mplies that the above construction of $g(z)$ furnishes the nalytic continuation of $g(z)$ into $D \cup \sigma \cup D^*$ (see [37], p.157).

Similarly $f(z)$ can be continued into $D \cup \sigma \cup D^*$. Equation 11.7) now gives $(z,\bar{z})$ for arbitrary z in $D \cup \sigma \cup D^*$ i.e. $J(z,\bar{z})$ has been extended from z in $D \cup \sigma \cup D^*$.

heorem 11.1 ([31]): Let $u(x,y) = \mathcal{U}(z,\bar{z}) \in C^2(D) \cap C^1(\bar{D})$ satisfy $[\mathcal{U}] = 0$ in D and suppose $\delta(z) = \mathcal{U}(z,z)$ is regular in $\cup \sigma \cup D^*$. Then $u(x,y)$ can be analytically continued into all f $D \cup \sigma \cup D^*$ where D^* is the mirror image of D reflected across .

2. The Envelope Method and Analytic Continuation

e have already seen one method of analytic continuation, that s in sections 3 and 4 where regularity of Cauchy data deter- ines regularity of solution. The following theorem shows how ntegral operators can also be used for this purpose.

Theorem 12.1. (The Envelope Method [21]): Let $F(z) \equiv F(z_1, z_2, \ldots, z_n)$ be defined by the integral representation

$$F(z) = \int_{\mathcal{L}} K(z;\zeta)d\zeta \qquad (12.1$$

where $K(z;\zeta)$ is a holomorphic function of (n+1) complex variables for $(z;\zeta)$ contained in, save for certain singularities indicated below, C^{n+1}. Furthermore, let the integration path $\mathcal{L}$ be a closed rectifiable contour, and let all of the singula points of $K(z;\zeta)$ be contained on the analytic set $\mathcal{G}_0 = \{z | S(z,\zeta) = 0; \zeta \in C^1\}$. Then $F(z)$ is regular for all point

$$z \notin \mathcal{G} \equiv \mathcal{G}_0 \cap \mathcal{G}_1$$

where

$$\mathcal{G}_1 = \{z | \frac{\partial S(z,\zeta)}{\partial \zeta} = 0; \ \zeta \in C^1\} \qquad (12.2$$

Proof: Let $F(z)$ be regular at $z = z_0$ and hence in a neighbou hood $N(z^0)$ of z^0. Now analytically continue $F(z)$ along a pat γ with one endpoint in $N(z^0)$. This can be done as long as no point of γ corresponds to a singularity of the integrand on $\mathcal{L}$ Even when this happens we can keep on continuing $F(z)$ along γ by deforming the path of integration to avoid the singularity $\zeta = \alpha(z)$ threatening to cross it. In particular suppose we have continued $F(z)$ along γ to a point $z = z_1$ and at that poi there exists a singularity $\zeta = \alpha$ on $\mathcal{L}$. Suppose however $S(z_1,$ has a simple zero at $\zeta = \alpha$, i.e. in a sufficiently small neig bourhood $N(\alpha) = \{\zeta | \ |\zeta - \alpha| < \epsilon\}$ we have

$$S(z_1,\zeta) \approx (\zeta - \alpha) \frac{\partial S(z_1;\alpha)}{\partial \zeta}$$

where $\dfrac{\partial S(z_1,\alpha)}{\partial \zeta} \neq 0$. Then we can deform $\mathcal{L}$ about the point $\zeta = \alpha$ by letting it follow a portion of the circle $|\zeta - \alpha| = \epsilon$ which implies that $F(z)$ is regular at z_1. Q.E.D.

Corollary 12.1 (Hadamard): Let n = 1 and suppose f(z) is singular at $\alpha_1, \alpha_2, \ldots$ and g(z) is singular at $\beta_1, \beta_2, \ldots$. Then

$$F(z) = \frac{1}{2\pi i}\int_{\mathcal{L}} f(\zeta)g(^z/_\zeta)^{d\zeta}/_\zeta \qquad (12.3)$$

is regular for $z \neq \alpha m\ \beta_n$, m,n = 1, 2, ...

Proof: Without loss of generality suppose that α_m and β_n are the only finite singularities of f(z) and g(z) respectively. Then

$$S(z,\zeta) = (\zeta - \alpha_m)(z - \beta_n\zeta)$$

and

$$\frac{\partial S(z,\zeta)}{\partial\zeta} = z - 2\beta_n\zeta + \alpha_m\beta_n$$

setting $S(z,\zeta) = \frac{\partial S(z,\zeta)}{\partial\zeta} = 0$ and eliminating ζ gives $(z - \alpha_m\beta_n)^2 = 0$, which implies the corollary.

Remark 1: If $\mathcal{L}$ is not a closed contour but an open contour between two fixed points ζ_1 and ζ_2, then we cannot deform $\mathcal{L}$ away from these points, and hence F(z) may be singular on the set

$$\mathcal{G}_3 = \{z | S(z,\zeta) = 0; \quad \zeta = \zeta_1 \text{ and } \zeta = \zeta_2\}. \qquad (12.4)$$

Such singularities are called endpoint-pinch singularities. In summary the possible singularities are those points which we are unable to list as regular points by the Hadamard method or envelope method (taking into account possible endpoint-pinch singularities).

Remark 2: For extensions of Theorem 12.1 and Corollary 12.1 to multiple integrals see Chapter 1 of [21]. Theorem 12.1 is often mistakenly credited to Landau, Bjorken and/or Polkinghorne and Screaton. It was actually first proved by R.P. Gilbert in his 1958 thesis. For a historical discussion of the origins of Theorem 12.1 see the introduction in [21].

The envelope method can be applied whenever an integral representation of the solution is available, e.g. the representation derived in sections 6-10. Here we apply Theorem 12.1 to the axially symmetric potential equation, i.e. the equation

$$\frac{\partial^2 u}{\partial z^2} + \frac{\partial^2 u}{\partial r^2} + \frac{1}{r}\frac{\partial u}{\partial r} = 0 \qquad (12.5)$$

where $(r.z.\phi)$ are cylindrical coordinates and u is assumed to be independent of ϕ.

Exercise 12.1: Show that if $u(z,r)$ is an analytic solution of (12.5) in some neighbourhood of the origin, then $u(z,r)$ is an even function of r and is uniquely determined by $u(z,0) = f(z)$.

Exercise 12.2: Using Theorem 9.2 show that for every analytic function $f(z)$

$$u(z,r) = \underset{\sim}{A}\{f\} \equiv \frac{1}{2\pi i}\int_{\mathcal{L}} f(\sigma)\,\frac{d\zeta}{\zeta} \qquad (12.6)$$

where

$$\mathcal{L} = \{\zeta \mid \zeta = e^{i\phi}\ ;\ 0 \le \phi \le 2\pi\} \qquad (12.7)$$

$$\sigma = z + i\frac{r}{2}\,(\zeta + \zeta^{-1}) \qquad (12.8)$$

defines a regular solution of (12.5) in some neighbourhood of the origin such that $u(z,0) = f(z)$.

Theorem 12.2: If the only finite singularities of $f(\sigma)$ are at $\sigma = \alpha$ then the only possible singularities of $u(z,r) = \mathcal{U}(\eta,\bar{\eta})(\eta = z+ir,\ \bar{\eta} = z-ir)$ on its first Riemann sheet are at $\eta = \alpha$ and $\eta = \bar{\alpha}$.

Proof: We represent $u(z,r)$ by the operator $\underset{\sim}{A}$ and apply the enevelope method. "Envelope" singularities:

$$\mathfrak{S}_0 = \{(z,r) \mid S(z,r;\zeta) = (z - \alpha)\zeta + \frac{ir}{2}(\zeta^2 + 1) = 0\}$$

$$\mathcal{G}_1 = \{(z,r) \mid \frac{\partial S(z,r;\zeta)}{\partial \zeta} = (z - \alpha) + ir\zeta = 0\}.$$

liminating ζ gives

$$\mathcal{G} = \mathcal{G}_0 \cap \mathcal{G}_1 = \{(z,r) \mid (z - \alpha)^2 + r^2 = 0\}$$
$$= \{\eta \mid (\eta - \alpha)(\eta - \bar{\alpha}) = 0\},$$

here $\eta = z + ir$.

Hadamard" singularities:

$$\mathcal{G}_0 \cap \{\zeta = 0\} = \{(z,r) \mid r = 0\}.$$

But it can easily be verified by termwise integration in quation 12.6 that $\underset{\sim}{A}\{f\}$ is regular about origin if $f(\sigma)$ is. But 12.5) is invariant under translations along z axis and hence $\{f\}$ is regular at points on the z axis provided they correspond to regular points of $f(\sigma)$.

emarks: Theorem 12.2 has a long history: [15], [25], [22] nd [11]. For further applications of Theorem 12.1 to the nalytic continuation of solutions of partial differential quations see [21].

3. The Axially Symmetric Helmholtz Equation

onsider

$$\Delta_3 u + u = 0 \tag{13.1}$$

efined in the exterior of a bounded domain. Assume that in ylindrical coordinates (r,z,ϕ) u is independent of ϕ. Then 13.1) becomes

$$L[u] \equiv u_{zz} + u_{rr} + \frac{1}{r} u_r + u = 0. \tag{13.2}$$

ow let D be a bounded, simply connected domain in the (r,z) lane which is symmetric with respect to the axis r = 0 and has mooth boundary ∂D. Let f(r,z) be a continuous function defined on $\partial D^+ = \partial D \cap \{(r,z) | r \geq 0\}$

$$u\Big|_{\partial D^+} = f \qquad (13.$$

Example 13.1: $\frac{\sin R}{R}$ (where $z = R\cos\theta$, $r = R \sin \theta$) is a solution of (13.2) which vanishes on the circle $R = \pi$, i.e. uniqueness does not in general hold for the boundary value problem (13.2), (13.3).

Theorem 13.1 ([40]): There is at most one solution of (13.2) (13.3) which is regular in the exterior of D and satisfies

$$\lim_{R\to\infty} R \left(\frac{\partial u}{\partial R} - iu\right) = 0 \qquad (13.$$

uniformly for $\theta \in [0,\pi]$.

Remark 1: (13.4) is known as the Sommerfeld radiation condition.

Remark 2: It is easily shown (c.f. Exercise 12.1) that the regularity of u in the exterior of D implies that u is an even function of r.

Proof of Theorem: On the circle of radius R (where the radiu of D is less than R) expand u in a Legendre series

$$u(R,\theta) = \sum_{n=0}^{\infty} a_n(R)P_n(\cos\theta); \; R > R_0 \qquad (13.$$

where

$$a_n(R) = \int_0^\pi u(R,\theta)P_n(\cos\theta) \sin \theta \, d\theta. \qquad (13.$$

From (13.2) and (13.6) it can be verified that

$$a_n(R) = a_n \; R^{-\frac{1}{2}}H^{(1)}_{n+\frac{1}{2}} (R) + b_n \; R^{-\frac{1}{2}}H^{(2)}_{n+\frac{1}{2}} (R) \qquad (13.$$

where $H^{(i)}_{n+\frac{1}{2}}$ denotes the Hankel function of the ith kind.

But (c.f. [16])

$$H^{(1)}_{n+\frac{1}{2}}(R) = \sqrt{\frac{2}{\pi R}}\, e^{i(R - \frac{1}{2}n\pi - \frac{1}{2}\pi)} + O\left(R^{-3/2}\right); \quad R \to \infty$$

$$H^{(2)}_{n+\frac{1}{2}}(R) = \sqrt{\frac{2}{\pi R}}\, e^{-i(R - \frac{1}{2}n\pi - \frac{1}{2}\pi)} + O\left(R^{-3/2}\right); \quad R \to \infty \qquad (13.8)$$

which implies (by the Sommerfeld radiation condition) that $b_n = 0$. Hence

$$u(R,\theta) - R^{-\frac{1}{2}} \sum_{n=0}^{\infty} a_n H^{(1)}_{n+\frac{1}{2}}(R) P_n(\cos\theta); \quad R > R_0. \qquad (13.9)$$

Now assume $u = 0$ on ∂D^+ and apply Green's formula in region B below to u and $\overline{u}$:

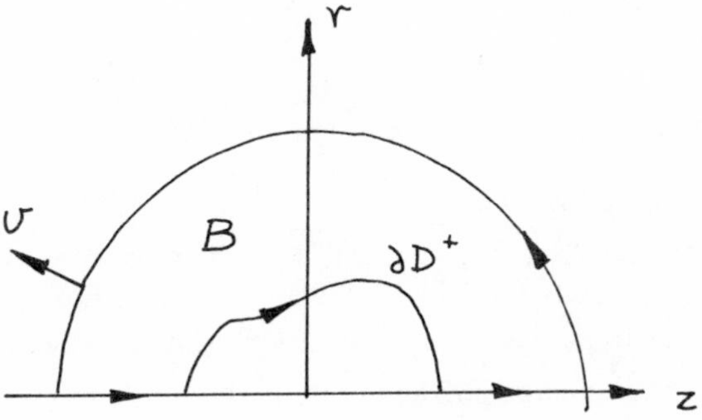

figure 13.1

$$\int_{\partial B} r\left(u \frac{\partial \overline{u}}{\partial v} - \overline{u}\frac{\partial u}{\partial v}\right) dS = \iint_B r(uM(\overline{u}) - \overline{u}M(u))\,dV \qquad (13.10)$$

here $M(u) = u_{zz} + u_{rr} + \frac{1}{r} u_r$, the axially symmetric harmonic quation. But (13.10) implies (since $L(u) = L(\overline{u}) = 0$) that

$$\int_0^{\pi} R^2 \left(u \frac{\partial \overline{u}}{\partial R} - \overline{u}\frac{\partial u}{\partial R}\right) \sin\, d\theta = 0. \qquad (13.11)$$

rom the relations (see [16])

$$\lim_{n\to\infty} \frac{(R/_2)^{n+\frac{1}{2}}}{\Gamma(n+\frac{1}{2})} \quad H^{(1)}_{n+\frac{1}{2}}(R) = -\frac{i}{\pi} \tag{13.1}$$

$$\max_{\theta\in[0,\pi]} |P_n(\cos\theta)| \leq 1 ,$$

it is seen that the series (13.9) can be differentiated term-wise. Furthermore from Abel's formula for the Wronskian we h (using equation (13.8) to evaluate the constant $-\frac{4i}{\pi}$)

$$H^{(1)}_{n+\frac{1}{2}}(R) \frac{d}{dR} H^{(2)}_{n+\frac{1}{2}}(R) - H^{(2)}_{n+\frac{1}{2}}(R) \frac{d}{dR} H^{(1)}_{n+\frac{1}{2}}(R) = \frac{-4i}{\pi R} \tag{13.1}$$

Since $\overline{H^{(1)}_{n+\frac{1}{2}}} = H^{(2)}_{n+\frac{1}{2}}$, substitution of (13.9) into (13.11) and making use of the orthogonality property of the $P_n(\cos\theta)$, giv

$$0 = \int_0^\pi R^2 (u\frac{\partial\bar{u}}{\partial R} - \bar{u}\frac{\partial u}{\partial R}) \sin\theta \, d\theta = -\frac{4i}{\pi} \Sigma |a_n|^2. \tag{13.1}$$

Hence $a_n = 0$ for every n, which implies that $u \equiv 0$.

<u>Theorem 13.2 ([29],[40])</u>: Let $u(R,\theta)$ be a regular solution o (13.2) for $R > C$ satisfying the Sommerfeld radiation conditio Then $u(R,\theta)$ has the representation

$$u(R,\theta) = R^{-\frac{1}{2}} H^{(1)}_{\frac{1}{2}}(R) \sum_{n=0}^{\infty} \frac{F_n(\cos\theta)}{R^n} + R^{-\frac{1}{2}} H^{(1)}_{3/2}(R) \sum_{n=0}^{\infty} \frac{G_n(\cos\theta)}{R^n}$$

(13.15

where the series converge uniformly and absolutely for $R \geq C > C$, $\theta \in [0,\pi]$. $F_0(\cos\theta)$ and $G_0(\cos\theta)$ determine $u(R,\theta$ uniquely.

<u>Proof</u>: If the recursion formula

$$H_{v-1}(R) + H_{v+1}(R) = \frac{2v}{R} H_v(R); \tag{13.16}$$

is used repeatedly one gets

$$H^{(1)}_{n+\frac{1}{2}}(R) = H^{(1)}_{\frac{1}{2}}(R)\, R_{n,\frac{1}{2}}(R) - H_{-\frac{1}{2}}(R)\, R_{n-1,3/2}(R) \qquad (13.17)$$

here $R_{n,\nu}(R)$ are Lommel polynomials defined as

$$R_{n,\nu}(z) = \sum_{k=0}^{[n/2]} \frac{(-1)^n (n-k)!\,\Gamma(\nu+n-k)}{k!\,(n-2k)!\,\Gamma(\nu+k)} \left(\frac{z}{2}\right)^{-n+2k} \qquad (13.18)$$

onsider now the series ($z = R$, $\xi = \cos\theta$)

$$E(z,\xi) = \sum_{n=0}^{\infty} a_n R_{n,\frac{1}{2}}(z) P_n(\xi)$$

$$Q(z,\xi) = \sum_{n=0}^{\infty} a_n R_{n-1,3/2}(z) P_n(\xi) \qquad (13.19)$$

hich result when (13.17) is substituted into the series rep-
esentation (13.9) of $u(R,\theta)$. Now set $z = C'e^{i\alpha}$, $0 \leq \alpha \leq 2\pi$.
e can rewrite $E(z,\xi)$ as

$$E(C'e^{i\alpha},\xi) = \sum_{n=0}^{\infty} \frac{a_n P_n(\xi) \frac{1}{\Gamma(n+\frac{1}{2})} \left(\frac{C'}{2} e^{i\alpha}\right)^{n+\frac{1}{2}} H^{(1)}_{n+\frac{1}{2}}(C')}{\frac{1}{\Gamma(n+\frac{1}{2})} \left(\frac{C'}{2} e^{i\alpha}\right)^{n+\frac{1}{2}} H^{(1)}_{n+\frac{1}{2}}(C')} R_{n,\frac{1}{2}}(C'e^{i\alpha}) \qquad (13.20)$$

From [16], p.35 we have

$$\lim_{n\to\infty} \frac{(z/2)^{n+\frac{1}{2}} R_{n,\frac{1}{2}}(z)}{\Gamma(n+\frac{1}{2})} = \left(\frac{z}{2}\right) J_{\frac{1}{2}}(z). \qquad (13.21)$$

Now note that (c.f. equation (13.12))

$$C'^{-\frac{1}{2}} \sum_{n=0}^{\infty} a_n H^{(1)}_{n+\frac{1}{2}}(C') P_n(\xi) \qquad (13.22)$$

is absolutely and uniformly convergent for $\xi\in[-1,+1]$. Equation (13.12), (13.20) and (13.21) imply that $E(z,\xi)$ is uniformly and absolutely convergent for $|z| = C'$, $\xi\in[-1, +1]$. But the series defining $E(z,\xi)$ is a series of polynomials in $1/z$ which converges uniformly and absolutely on the circle $|1/z| = 1/C'$ and hence is analytic for $|1/z| < \frac{1}{C'}$, $\xi\in[-1, +1]$. A similar result holds for $Q(z,\xi)$. Hence

$$u(R,\theta) = R^{-\frac{1}{2}}H_{\frac{1}{2}}^{(1)}(R)E(\tfrac{1}{R}, \cos\theta) + R^{-\frac{1}{2}}H_{3/2}^{(1)}(R)Q(\tfrac{1}{R}, \cos\theta) \tag{13.23}$$

where E and Q are analytic functions of $\frac{1}{R}$ and are regular in the interior of the circle $|\frac{1}{R}|= \frac{1}{C'}$ in the complex $\frac{1}{R}$ plane. Equation (13.15) follows from this statement. If F_0 and G_0 are known, F_n and G_n can be computed recursively by substituti[ng] into equation (13.2). It is easily verified that

$$F_0(\cos\theta) = \sum_{n=0}^{\infty} a_{2n} (-1)^n P_{2n}(\cos\theta)$$

$$G_0(\cos\theta) = \sum_{n=0}^{\infty} a_{2n+1}(-1)^n P_{2n+1}(\cos\theta). \tag{13.24}$$

Corollary ([29], [40]): Let $u(R,\theta)$ be a solution of (13.2) f[or] $R > R_0$, $\theta\in[0,\pi]$ satisfying the Sommerfeld radiation condition. Then

$$\lim_{R\to\infty} e^{-iR} Ru(R,\theta) = f(\cos\theta) = -G_0(\cos\theta) -iF_0(\cos\theta)$$

exists uniformly for $\theta\in[0,\pi]$. If $u(R,\theta)$ has the expansion (f[or] $R > R_0$, $\theta\in[0,\pi]$)

$$u(R,\theta) = \sqrt{\frac{\pi}{2}}\, R^{-\frac{1}{2}} \sum_{n=0}^{\infty} a_n i^{n+1} H_{n+\frac{1}{2}}^{(1)}(R)P_n(\cos\theta) \tag{13.25}$$

then

$$f(\cos\theta) = \sum_{n=0}^{\infty} a_n P_n(\cos\theta) \tag{13.26}$$

Proof: The corollary follows from equations(13.8), (13.15) and (13.24).

We now consider the inverse scattering problem associated with equation (13.2), i.e. given the radiation pattern $f(\cos\theta)$, to determine $u(R,\theta)$ and its domain of regularity. In particular we want to analytically continue $u(R,\theta)$, from its initial domain of definition in a neighbourhood of infinity.

14. Analytic Continuation of Solutions to the Axially Symmetric Helmholtz Equation

Let u be a regular solution of

$$u_{zz} + u_{rr} + \frac{1}{r}u_r + u = 0 \tag{14.1}$$

in the exterior of a bounded domain D, and suppose

$$\lim_{R\to\infty} R\left(\frac{\partial u}{\partial R} - iu\right) = 0; \quad R = \sqrt{r^2 + z^2} \tag{14.2}$$

From section 13 we have

$$u \sim \frac{e^{iR}}{R} f(\cos\theta); \quad R \to \infty . \tag{14.3}$$

where $f(\cos\theta)$ is known as the radiation pattern of u.

Theorem 14.1 ([33]): A necessary and sufficient condition for a function $f(\cos\theta)$ to be a radiation pattern is that there exists an (axially symmetric) harmonic function $h(z,r) = \tilde{h}(R,\theta)$ which is regular in the entire space such that $h(1,\theta) = f(\cos\theta)$ and furthermore has the property that

$$\int_0^{\pi} |\tilde{h}(R,\theta)|^2 \sin\theta \, d\theta \tag{14.4}$$

is an entire function of R of order one and finite type C. When this condition holds there exists a unique function $u(z,r) = \tilde{u}(R,\theta)$ which satisfies the Sommerfeld radiation condition (14.2) and is a regular solution of the (axially symmetric) Helmholtz equation for $R > C$ such that

$$\tilde{u}(R,\theta) \sim \frac{e^{iR}}{R} f(\cos\theta) + O\left(\frac{1}{R^2}\right); \ R \to \infty. \tag{14.5}$$

<u>Proof</u>: Suppose $f(\cos\theta)$ is a radiation pattern. Then

$$f(\cos\theta) = \sum_{n=0}^{\infty} a_n P_n(\cos\theta) \tag{14.6}$$

where the series

$$\sum_{n=0}^{\infty} a_n \ i^{n+1} \ H^{(1)}_{n+\frac{1}{2}}(C')P_n(\cos\theta) \tag{14.7}$$

converges absolutely and uniformly for $C' > C$ for some $C > 0$. From equation (13.12) this implies that

$$|a_n|\Gamma(n + \tfrac{1}{2})(2/C')^{n+\frac{1}{2}} \tag{14.8}$$

is bounded, i.e. (using Stirling's formula)

$$\overline{\lim_{n\to\infty}} \ n|a_n|^{1/n} = \tfrac{1}{2} eC'. \tag{14.9}$$

But $\tilde{h}(R,\theta) = \sum_{n=0}^{\infty} a_n R^n P_n(\cos\theta)$ and (14.9) implies that

$\int_0^{\pi} |\tilde{h}(R,\theta)|^2 \sin\theta d\theta = \sum_{n=0}^{\infty} \frac{2}{2n+1} |a_n|2R^{2n}$ is an entire function of order 1 and exponential type C.

Suppose $\int_0^{\pi} |\tilde{h}(R,\theta)|^2 \sin\theta d\theta = \sum_{n=0}^{\infty} \frac{2}{2n+1} |a_n|2R^{2n}$ is an entire function of order 1 and exponential type C. Then from equations (14.8) and (14.9) the series (14.7) converges for each $C' > C$.

From equation (13.12), the series

$$\tilde{u}(R,\theta) = \sqrt{\frac{\pi}{2}} \ R^{-\frac{1}{2}} \sum_{n=0}^{\infty} a_n i^{n+1} H^{(1)}_{n+\frac{1}{2}} (R)P_n(\cos\theta) \tag{14.10}$$

can be differentiated termwise and defines a solution of (14.1) for $R > C$ which satisfies the Sommerfeld radiation condition. From the corollary to Theorem 13.2 $u(r,\theta)$ has $f(\cos\theta)$ as its radiation pattern.

We now want to analytically continue $\tilde{u}(R,\theta)$ pastthe circle $R=C$. From section 12 $h(z,r)$ is uniquely determined by the function

$$h(z,0) = \sum_{n=0}^{\infty} a_n z^n = \tfrac{1}{2}\int_{-1}^{+1} f(\xi)\,\frac{1-z^2}{(1-2\xi z+z^2)^{3/2}}d\xi.$$

Equation (14.9) implies that $h(z,0)$ is an entire function of order 1 and type $^C/2$. Let $f(z)$ be the Borel transform of $h(2iz,0)$, i.e.

$$f(z) = \sum_{n=0}^{\infty} a_n 2^n i^n n!\; z^{-n-1}. \tag{14.11}$$

Before we can state our first lemma we will need to introduce the concept of the indicator diagram of an entire function of exponential type. Suppose $g(z)$ is an entire function of exponential type. Then the indicator function of $g(z)$ is defined as

$$k(\theta) = \overline{\lim_{R\to\infty}}\; R^{-1}\log|g(Re^{i\theta})|.$$

It can be shown that $k(\theta)$ is the supporting functio n of a convex set, called the indicator diagram of $g(z)$.

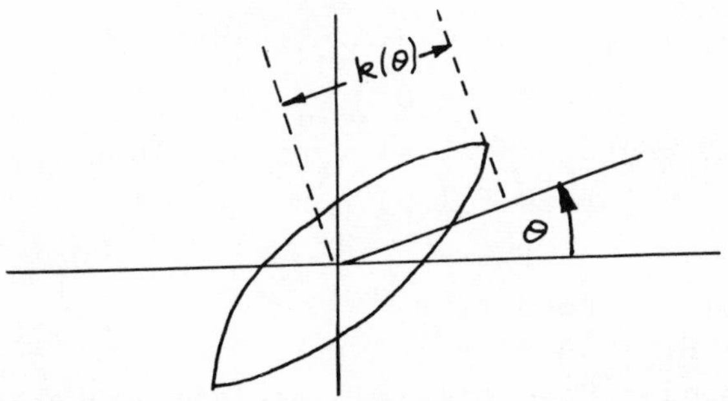

figure 14.1

Lemma 14.1 (Polya): $f(z)$ is regular in the exterior of the c jugate indicator diagram $\mathcal{G}$ of $h(2iz,0)$.

Proof: [4], p.75.

Remark: $\mathcal{G}$ is a closed convex set contained in $|z| \leq C$. Now define

$$g(z) = C'^{-\frac{1}{2}} \sqrt{\frac{\pi}{2}} \sum_{n=0}^{\infty} a_n i^{n+1} H^{(1)}_{n+\frac{1}{2}}(C')\left(\frac{z}{C'}\right)^{-n-1} \qquad (14.$$

where $C' > C$.

Lemma 14.2 ([12]): $g(z)$ is regular in the exterior of $\mathcal{G}$.

Proof: From Lemma 14.1 and Hadamard's multiplication of singularities theorem (Corollary 12.1; [37], p. 157) it suffi ces to show that the singularities of

$$G(z) = \sum_{n=0}^{\infty} \frac{H^{(1)}_{n+\frac{1}{2}}(C')}{n!\ 2^n} \left(\frac{z}{C'}\right)^{-n} \qquad (14.$$

lie on the closed interval [0,1]. But from [16] p.78, 100 w can actually sum (14.13) to give

$$G(z) = -i\sqrt{\frac{2}{\pi C'}}\,\left(1 - \frac{1}{z}\right)^{-\frac{1}{2}} e^{iC'(1 - {}^1/z)^{\frac{1}{2}}} \qquad (14.1$$

i.e. the only singularities of $G(z)$ are branch points at $z=0$ and $z = +1$.

We now construct the axially symmetric harmonic function $v(z,r)$ such that $v(z,0) = g(z)$:

$$v(z,r) = \tilde{v}(R,\theta) = \sqrt{\frac{\pi}{2}}\, C'^{-\frac{1}{2}} \sum_{n=0}^{\infty} a_n i^{n+1} H^{(1)}_{n+\frac{1}{2}}(C')\left(\frac{R}{C'}\right)^{-n-1} P_n(\cos\theta$$

(14.1

Note that $\frac{1}{R}\tilde{v}(\frac{1}{R},\theta)$ is an axially symmetric harmonic functio in a neighbourhood of the origin. Applying Theorem 12.2 to $\frac{1}{R}\tilde{v}(\frac{1}{R},\theta)$ and using Lemma 14.2 we have that $v(z,r) = \tilde{v}(R,\theta)$ is regular in the exterior of $G \cup \overline{G}$ (where $\overline{G}$ denotes the image of G under conjugation). By the law of permanence of functional equations $v(z,r)$ is harmonic in $G \cup \overline{G}$. By construction we have that $v(C',\theta) = \tilde{u}(C',\theta)$.

<u>Lemma 14.3 ([12])</u>: Let $A = \{(z,r) | \sqrt{z^2 + r^2} = C'\}$. Then for $(z,r) \in A$, $v(z,r)$ $(= u(z,r))$ is an analytic function of $\eta = z + ir$ and can be continued analytically as a function of η into the exterior of $G \cup \overline{G}$ $\{(r,z) | r = 0\}$.

<u>Proof</u>: Let ϕ conformally map the exterior of the circle A onto the upper plane in the complex $\zeta = x + iy$ plane such that ϕ maps the line $r = 0$ onto the line $x = 0$. Under such a mapping the exterior of $G \cup \overline{G}$ is taken into a region Ω containing the upper half plane in its interior. The equation for $v(z,r)$ is transformed into an elliptic equation for a function $w(x,y)$ with coefficients analytic in $\Omega' = \Omega - \{(x,y) | x = 0\}$. Since $v(z,r)$ is regular in the exterior of $G \cup \overline{G}$, by Theorem B $w(x,y) = W(\zeta,\overline{\zeta})$ is an analytic function of ζ and $\overline{\zeta}$ for $(\zeta,\overline{\zeta}) \in \Omega' \times \overline{\Omega}'$ where $\overline{\Omega}'$is the image of Ω' under conjugation. Hence we can conclude that

$$f(\zeta) = W(\zeta,\zeta) \tag{14.16}$$

is regular for ζ in $\Omega' \cap \overline{\Omega}'$. Using the inverse conformal mapping now shows that $v(z,r)\big|_{(z,r)\in A}$ can be continued to an analytic function of η in the inverse image of $\Omega' \cap \overline{\Omega}'$. From Theorem 11.1 we can now conclude that $u(z,r)$ is regular in the exterior of $G \cup \overline{G} \cup \{(r,z) | r = 0\}$.

<u>Remark</u>: A similar analysis shows that $\frac{\partial u}{\partial R}\Big|_{R=C'}$ can be continued to an analytic function of η in the exterior of $G \cup \overline{G} \cup \{r = 0\}$. Hence the continuation of $u(z,r)$ across the circle $R = C'$ can be accomplished by referring to the results of section 4 instead of making use of Lewy's reflection principle.

We now collect our results in the following theorem:

Theorem 14.2 ([12]): Let $f(\cos\theta)$ be the radiation pattern of a solution $\tilde{u}(R,\theta)$ of the three dimensional axially symmetric Helmholtz equation where (R,θ) are polar coordinates and let

$$F(z) = \tfrac{1}{2}\int_{-1}^{+1} f(\xi)\ \frac{(1 + 4z^2)d\xi}{(1 - 4iz\xi - 4z^2)^{3/2}}.$$

The $F(z)$, $z = Re^{i\theta}$, is an entire function of order one and finite exponential type C. If G is the conjugate indicator diagram of $F(z)$, then $\tilde{u}(R,\ )$ is regular in the exterior of $G \cup \overline{G} \cup \{(R,\theta)\,|\,\theta = 0,\pi, R \leq C\}$ (see figure 14.2 below).

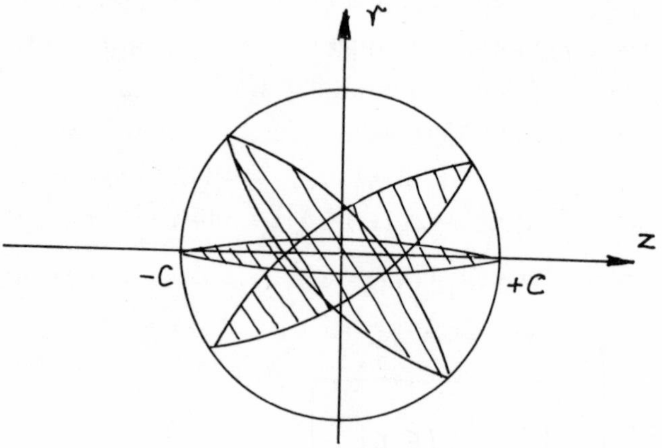

figure 14.2

Remark: The results of this section have recently been extended by B.D. Sleeman in

B.D. Sleeman, The three-dimensional inverse scattering problem for the Helmholtz equation, Proc. Camb. Phil. Soc. 73 (1973), 477-488.

Pseudoparabolic equations

5. <u>Pseudoparabolic Equations in One Space Variable</u>

onsider the pseudoparabolic equation (c.f. [36])

$$\mathcal{L}[u] \equiv u_{xtx} + d(x,t)u_t + \eta u_{xx} + a(x)u_x + b(x)u = q(x,t) \tag{15.1}$$

efined in $D(H,T) = \{(x,t)\,|\,0 < x < H,\ 0 < t < T\}$. Assume $(x,t) \in C^1\ (\overline{D}(H,T))$ ($\overline{D}(H,T)$ denotes the closure of $D(H,T)$), $(x,t) \in C^0\ (\overline{D}(H,T))$, $a(x) \in C^1[0,H]$ and $b(x) \in C^0[0,H]$. η is a onstant. (References for the appearance of equation (15.1) n physics can be found in [5] and [36]).

Define the adjoint equation by

$$\mathcal{M}[v] = v_{xtx} + d(x,t)v_t - \eta v_{xx} + (av)_x - bv = 0. \tag{15.2}$$

Now let $(\xi,\tau) \in D(H,T)$ and integrate

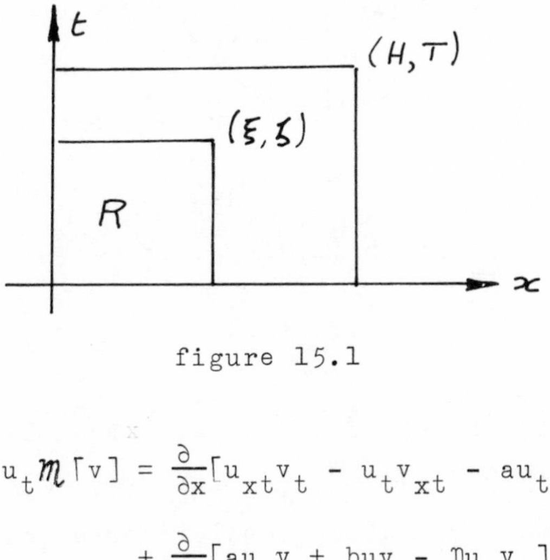

figure 15.1

ne identity

$$v_t\mathcal{L}[u] - u_t\mathcal{M}[v] = \frac{\partial}{\partial x}[u_{xt}v_t - u_tv_{xt} - au_tv + nu_xv_t + \eta u_tv_x]$$

$$+ \frac{\partial}{\partial t}[au_xv + buv - \eta u_xv_x]$$

ver the rectangle R in figure 15.1. An application of Green's formula gives

$$\int_0^{\tau}\int_0^{\xi}(v_t\mathcal{L}[u] - u_t\mathcal{M}[v])dxdt = \tag{15.3}$$

$$= \int_{\partial R}(u_{xt}v_t - u_t v_{xt} - au_t v + \eta u_x v_t + \eta u_t v_x)dt - (au_x v + buv - \eta u_x v_x)dx$$

Suppose there exists a function $v(x,t;\xi,\tau)$ such that

$$\mathcal{M}[v] = 0 \tag{15.4}$$

$$v_x(\xi,t;\xi,\tau) = \frac{1}{\eta}[1 - e^{\eta(t-\tau)}] \tag{15.4}$$

$$v(\xi,t;\xi,\tau) = 0 \tag{15.4}$$

$$v(x,\tau;\xi,\tau) = 0. \tag{15.4}$$

If $\eta = 0$, (15.4b) is interpreted in its limiting form as $n \to 0$. Then if there exists a function $u(x,t)$ such that

$$\mathcal{L}[u] = q \tag{15.5}$$

$$u(0,t) = f(t) \tag{15.5}$$

$$u_x(0,t) = g(t) \tag{15.5}$$

$$u(x,0) = h(x) \tag{15.5}$$

where $f(t),\ g(t) \in C^1[0,T]$, $h(x) \in C^2[0,H]$, then equation (15.3) implies that

$$\begin{aligned} u(\xi,\tau) = h(\xi) &+ \int_0^{\xi}[a(x)h'(x)v(x,0;\xi,\tau) - \eta h'(x)v_x(x,0;\xi,\tau) \\ &\qquad + b(x)h(x)v(x,0;\xi,\tau)] \\ &+ \int_0^{\tau}[g'(t)v_t(0,t;\xi,\tau) - f'(t)v_{xt}(0,t;\xi,\tau) \\ &\qquad -a(0)f'(t)v(0,t;\xi,\tau) + \eta g(t)v_t(0,t;\xi,\tau) \\ &\qquad + \eta f'(t)v_x(0,t;\xi,\tau)]dt \\ &+ \int_0^{\tau}\int_0^{\xi} q(x,t)v_t(x,t;\xi,\tau)dxdt. \end{aligned} \tag{15.6}$$

Equation (15.6) gives the solution of the Goursat problem (15.5) in terms of the Riemann function $v(x,t;\xi,\tau)$.

Exercise 15.1: Show $v(x,t;\xi,\tau)$ exists by using the methods of section 3. Let

$$||.||_\lambda = \max_{(x,t)\in R} \{e^{-\lambda[(\xi-x)+(\tau-t)]}|.|\}.$$

Show that as a function of ξ and τ, $\mathcal{L}[v] = 0$.

Exercise 15.2: Suppose the coefficients a, b and d are entire functions of x and q = 0. Show that $v(x,t;\xi,\tau)$ is an entire function of ξ. Conclude from equation (15.6) that if $h(x) = 0$ and $u(x,t)$ is a solution of $\mathcal{L}[u] = 0$ which is analytic in a neighbourhood of the origin, then $u(x,t)$ can be analytically continued into a strip of the form $|t| < t_0$, $-\infty < x < \infty$. Compare this result to the behaviour of solutions to parabolic and elliptic equations in two independent variables.

We now want so solve the first initial boundary value problem for equation (15.1) i.e. find a solution of $\mathcal{L}[u] = q$ in $D(H,T)$, continuously differentiable in $\overline{D}(H,T)$, such that

$$u(0,t) = f(t) \qquad (15.7a)$$

$$u(x,0) = h(x) \qquad (15.7b)$$

$$u(H,t) = \phi(t) \qquad (15.7c)$$

where $f(t)$, $\phi(t)\in C^1[0,T]$ $h(x)\in C^2[0,H]$. To find u set $\xi = H$ in (15.6) and integrate by parts to arrive at

$$\gamma(\tau) = g(\tau)v_t(0,\tau;H,\tau) + \int_0^\tau [\eta v_t(0,t;H,\tau) - v_{tt}(0,t;H,\tau) - \eta v_{xt}(0,t;H,\tau)]g(t)dt \qquad (15.8)$$

where

$$\begin{aligned}\gamma(\tau) = \phi(\tau) - h(H) &- \int_0^H [h'(x)(a(x)v(x,0;H,\tau) - \eta v_x(x,0;H,\tau\\
&+ h(x)b(x)v(x,0;H,\tau)]dx\\
&+ h'(0)[v_t(0,0;H,\tau) + \eta v_x(0,0;H,\tau)]\\
&+ \int_0^\tau f'(t)[v_{xt}(0,t;H,\tau) - a(0)v(0,t;H,\tau)\\
&+ \eta v_x(0,t;H,\tau)]dt\\
&- \int_0^\tau\int_0^H q(x,t)v_t(x,t;H,\tau)dxdt\end{aligned} \qquad (15.9$$

Exercise 15.3: Use exercise 15.1 and the fact that $d(x,t) \in C^1(\overline{D}(H,T))$ to show that $\gamma(\tau)$ and the kernel of the integral equation (15.8) is continuously differentiable with respe to τ for $0 \leq \tau \leq T$, $0 \leq t \leq \tau$; see also [20] pp. 116-117. Conclude that if a solution $g(\tau)$ exists then $g(\tau) \in C^1[0,T]$.

To show that a solution $g(\tau)$ of (15.8) exists, it suffices to show $v_t(0,\tau;H,\tau) \neq 0$ for $\tau \in [0,T]$. To this end consider

$$\mu(x) = v_t(x,\tau;H,\tau) \qquad (15.10$$

for arbitrary (but fixed) τ in $[0,T]$. The differential equation (15.4a) and the boundary condition (15.4d) imply that

$$\mu_{xx} + d(x,\tau)\mu = 0 \qquad (15.11$$

Suppose $d(x,t) \leq 0$ for $(x,t) \in \overline{D}(H,T)$. Then if $\mu(0) = 0$, $\mu(x) \equiv$ since (15.4c) implies that $\mu(H) = 0$. But $\mu(x) \equiv 0$ implies tha $u_x(H) = v_{xt}(H,\tau;H,\tau) = 0$ which contradicts (15.4b), ((15.4b) implies that $v_{xt}(H,\tau;H,\tau) = -1$). Hence if $d(x,t) \leq 0$ we can solve (15.8) for $g(\tau)$ and substitute into (15.6) to give the (unique) solution of the first initial boundary value problem for $\mathcal{L}[u] = q$.

eorem 15.1 ([13], [36]): Let d(x,t) be continuously ifferentiable and nonpositive in $\overline{D}(H,T)$, q(x,t) continuous n D(H,T), and assume $a(x)\in C'[0,H]$, $b(x)\in C[0,H]$. Let f(t), (t)$\in C^1[0,T]$ and $h(x)\in C^2[0,H]$. Then there exists a unique olution to $\mathcal{L}[u] = q(x,t)$ satisfying the initial - boundary ata (15.7).

xample 15.1: In general $d(x,t) \leq 0$ in $\overline{D}(H,T)$ is necessary. or example u(x,t) = t sin kx is a solution of

$$u_{xtx} + k^2u_t = 0 \tag{15.12}$$

or $(x,t)\in D(\frac{\pi}{k},T)$, T arbitrary. $u\in C^1[\overline{D}(\frac{\pi}{k},T)]$. But u(0,t) = $(\frac{\pi}{k},t)$ = u(x,0) = 0, i.e. the solution of the first initial oundary value problem is not unique.

A result similar to that of Theorem 15.1 can be obtained for seudoparabolic equations in an arbitrary number of space imensions. Rather than pursuing this investigation we will ow turn our attention to the analytic properties of pseudo-arabolic equations in two space variables.

6. Pseudoparabolic Equations in Two Space Variables.

onsider

$$\underset{\sim}{M}\left[\frac{\partial \underset{\sim}{u}}{\partial t}\right] + \gamma \underset{\sim}{L}[\underset{\sim}{u}] = 0 \tag{16.1}$$

here $\underset{\sim}{M} \equiv \Delta + \underset{\sim}{d}(x,y)$, $\underset{\sim}{L} = \Delta + \underset{\sim}{a}(x,y)\frac{\partial}{\partial x} + \underset{\sim}{b}(x,y)\frac{\partial}{\partial y} + \underset{\sim}{c}(x,y)$, nd γ is a constant. Let $\underset{\sim}{u} = e^{-\gamma t}u$. Then (16.1) becomes

$$\mathcal{L}[u] \equiv M\left[\frac{\partial u}{\partial t}\right] + L[u] = 0 \tag{16.2}$$

here

$$M = \Delta + d(x,y)$$

$$L = a(x,y)\frac{\partial}{\partial x} + b(x,y)\frac{\partial}{\partial y} + c(x,y).$$

Assumption: As a function of $z = x + iy$ and $z^* = x - iy$, a,b,c,d are analytic in $D \times D^*$ where $D^* = \{z^* | \overline{z^*} \in D\}$ and D is bounded simply connected domain in $\mathbb{R}^2$.
Define the adjoint of $\mathcal{L}[u] = 0$ to be

$$\mathfrak{M}[v] \equiv M[v_t] - L^*[v] = 0 \qquad (16.3)$$

where $L^*[v] = -(av)_x - (bv)_y + cv.$

Definition 16.1: A function S of the form

$$S(x,y,t;\xi,\eta,\tau) = A(x,y,t;\xi,\eta,\tau)\log\frac{1}{r} + B(x,y,t;\xi,\eta,\tau) \qquad (16.4)$$

where $r = [(x-\xi)^2 + (y-\eta)^2]^{\frac{1}{2}}$ will be called a fundamental solution if

1) $\mathfrak{M}[S] = 0$ for $r \neq 0$.

2) A and B are analytic functions of their independent variables.

3) $A_t = 1$ at $x = \xi$, $y = \eta$, $t = \tau$. $A = B = 0$ at $t = \tau$.

We will now construct S. Let

$$z = x + iy \qquad \zeta = \xi + i\eta$$
$$z^* = x - iy \qquad \zeta^* = \xi - i\eta \qquad (16.5)$$

Then (16.3) can be written as

$$\mathfrak{M}[V] = M[V_t] - L^*[V]$$
$$= V_{zz^*t} + \delta V_t + (\alpha V)_z + (\beta V)_{z^*} - \gamma V = 0 \qquad (16.6)$$

where $V(z,z^*,t) = v(x,y,t)$, $\alpha = \frac{1}{4}(a + ib)$, $\beta = \frac{1}{4}(a - ib)$, $\gamma = {}^c/_4$, $\delta = {}^d/_4$. Note that $r^2 = (z-\zeta)(z^*-\zeta^*)$.
Substituting (16.4) into (16.6) gives

$$\mathfrak{M}[S] = \mathfrak{M}[A]\log\frac{1}{r} - \frac{A_{zt} + \beta A}{2(z^*-\zeta^*)} - \frac{A_{z^*t} - \alpha A}{2(z-\zeta)} + \mathfrak{M}[B] = 0$$

which implies that

$$\mathfrak{M}[A] = 0 \tag{16.7}$$

$$\left[\frac{\partial^2}{\partial z\,\partial t} + \beta(z,\zeta^*)\right] A(z,\zeta^*,t;\zeta,\zeta^*,\tau) = 0 \tag{16.8}$$

$$\left[\frac{\partial^2}{\partial z^*\partial t} + \alpha(\zeta,z^*)\right] A(\zeta,z^*,t;\zeta,\zeta^*,\tau) = 0 \tag{16.9}$$

Once we have found A, B is any solution of

$$\mathfrak{M}[B] = \frac{A_{zt} + \beta A}{2(z^*-\zeta^*)} + \frac{A_{z^*t} + \alpha A}{2(z - \zeta)} . \tag{16.10}$$

Now let

$$A = \sum_{j=1}^{\infty} A_j(z,z^*;\zeta,\zeta^*) \frac{(t-\tau)^j}{j!} \tag{16.11}$$

Substituting (16.11) into equations (16.7) - (16.9) gives

$$\begin{aligned}
&\frac{\partial A_1}{\partial z} = 0 \text{ on } z^* = \zeta^* \\
&\frac{\partial A_1}{\partial z^*} = 0 \text{ on } z = \zeta \\
&\frac{\partial A_{j+1}}{\partial z} + \beta A_j = 0 \text{ on } z^* = \zeta^*;\ j = 0,1,2,\ldots \\
&\frac{\partial A_{j+1}}{\partial z^*} + \alpha A_j = 0 \text{ on } z = \zeta;\ j = 0,\ 1,2,\ \ldots
\end{aligned} \tag{16.12}$$

and

$$\begin{aligned}
&M[A_1] = 0 \\
&M[A_{j+1}] = L^*[A_j]
\end{aligned} \tag{16.13}$$

Condition (3) satisfied by S implies that

$$\begin{aligned}
&A_1(\zeta,\zeta^*;\zeta,\zeta^*) = 1 \\
&A_j(\zeta,\zeta^*;\zeta,\zeta^*) = 0,\ j = 2,\ 3,\ldots
\end{aligned} \tag{16.14}$$

and hence

$$A_1 = 1 \quad \text{on } z^* = \zeta^*$$

$$A_1 = 1 \quad \text{on } z = \zeta \tag{16.15}$$

and

$$A_{j+1}(z,\zeta^*;\zeta,\zeta^*) = -\int_{\zeta}^{z} \beta(\sigma,\zeta^*)A_j(\sigma,\zeta^*;\zeta,\zeta^*)d\sigma;\ j=1,2,\ldots$$

$$A_{j+1}(\zeta,z^*;\zeta,\zeta^*) = -\int_{\zeta^*}^{z^*} \alpha(\zeta,\rho)A_j(\zeta,\rho;\zeta,\zeta^*)d\rho;\ j = 1,2,\ldots \tag{16.16}$$

Note that equations (16.13) and (16.15) imply that A_1 is the Riemann function for $M[u] = 0$.

<u>Lemma 16.1 ([14]):</u> $A(z,z^*,t;\zeta,\zeta^*,\tau)$ is an analytic function of its six independent variables for all (complex) t, τ and z, $\zeta\in D$, z^*, $\zeta^*\in D^*$.

<u>Proof</u>: Integrating the identity (11.4) of section 11 with $U = A_{j+1}$, $V = A_1$, over the quadrilateral (ζ,ζ^*), (z,ζ^*), (z,z^*), (ζ,z^*) gives (after an integration by parts)

$$A_{j+1}(z,z^*;\zeta,\zeta^*) = -\int_{\zeta}^{z} A_1(\sigma,z^*;z,z^*)\beta(\sigma,z^*)A_j(\sigma,z^*;\zeta,\zeta^*)d\sigma$$

$$-\int_{\zeta^*}^{z^*} A_1(z,\rho,z,z^*)\alpha(z,\rho)A_j(z,\rho;\zeta,\zeta^*)d\rho$$

$$+\int_{\zeta}^{z}\int_{\zeta^*}^{z^*} A_1(\sigma,\rho;z,z^*)\gamma(\sigma,\rho) + \frac{\partial}{\partial\sigma}A_1(\sigma,\rho;z,z^*)\alpha(\sigma,\rho)$$

$$+ \frac{\partial}{\partial\rho}A_1(\sigma,\rho;z,z^*)\beta(\sigma,\rho)A_j(\sigma,\rho;\zeta,\zeta^*)d\rho d\sigma. \tag{16.17}$$

By induction A_j is analytic for z, $\zeta\in D$, z^*, $\zeta^*\in D^*$ (since the Riemann function A_1 is - see [20], p.141, and [39]). Let k be an upper bound on $|A_1\beta|$, $|A_1\alpha|$ and $|A_1\gamma + A_{1_z}\alpha + A_{1_{z^*}}\beta|$ for z, $\zeta\in\overline{\Omega}\subset D$, z^*, $\zeta^*\in\overline{\Omega}^* \subset D^*$, where $\overline{\Omega}$ and $\overline{\Omega}^*$ are arbitrary compact subsets of D and D* respectively. Let ℓ be an upper bound on the legnth of the paths of integration in (16.17) and let

$_1| \leq C$ for z, $\zeta\in\bar{\Omega}\subset D$, z^*, $\zeta^*\in\bar{\Omega}^* \subset D^*$. Then by induction 6.17) implies that

$$|A_j| \leq C\, k^j \ell^j (2 + \ell)^j; \; z,\ \zeta\in\Omega,\ z^*,\ \zeta^*\in\Omega^* \qquad (16.18)$$

d hence the series (16.11) converges in $|t - \tau| < T_1$, z, $\bar{\Omega}$, z^*, $\zeta^*\in\bar{\Omega}^*$, where T_1 is arbitrarily large. The lemma follows om this last statement.

Now look at the function B. Set

$$B = \sum_{j=1}^{\infty} B_j(z,z^*;\zeta,\zeta^*)\,\frac{(t-\tau)^j}{j!} \qquad (16.19)$$

bstituting (16.19) into (16.10) gives

$$M[B_{j+1}] = L^*[B_j] + \frac{\frac{\partial}{\partial z}A_{j+1} + \beta A_j}{2(z^* - \zeta^*)} + \frac{\frac{\partial}{\partial z^*}A_{j+1} + \alpha A_j}{2(z - \zeta)}; \; j=1,2,...$$

$$M[B_1] = \frac{\partial A_1}{\partial z}\Big/ 2(z^* - \zeta^*) + \frac{\partial A_1}{\partial z^*}\Big/ 2(z - \zeta) \qquad (16.20)$$

nce B is an arbitrary solution of (16.10), without loss of nerality we impose the boundary conditions

$$B_j(z,\zeta^*;\zeta,\zeta^*) = B_j(\zeta,z^*;\zeta,\zeta^*) = 0; \; j=1,2,3,... \qquad (16.21)$$

itating the proof of lemma 16.1 now gives

<u>mma 16.2 ([14])</u>: $B(z,z^*,t;\zeta,\zeta^*,\tau)$ is an analytic function of s six independent variables for all (complex) t, τ, and z, D, z^*, $\zeta^* \in D^*$.

w let

$$T = \{t \mid 0 \leq t < T_0\}$$

$$G(D\times T) = \{u(x,y,t) \mid u, u_x, u_y \in C^1(D\times T); u_{xyt}, u_{xxt}, u_{yyt} \in C^0(D\times T)\}$$

<u>eorem 16.1 ([14])</u>: Let $u(x,y,t)\in G(D\times T)$ be a solution of 6.2) in $D\times T$ and assume $\mathcal{U}(z,z^*,0) = u(x,y,0)$ is analytic in D^*. Then for each fixed $t\in T$, $\mathcal{U}(z,z^*,t) = u(x,y,t)$ is

an analytic function of z and z* in $D \times D^*$.

Proof: We first show that without loss of generality we can assume $\mathcal{U}(z,z^*,0) = 0$. Let $f(z,z^*) = L[\mathcal{U}(z,z^*,0)]$ and define

$$C(z,z^*,\zeta,\zeta^*,t) = \sum_{j=1}^{\infty} C_j(z,z^*,\zeta,\zeta^*) \frac{t^j}{j!} \tag{16.22}$$

where

$$M[C_1] = f(z,z^*)$$

$$M[C_{j+1}] = -L[C_j];\ j = 1, 2, \ldots \tag{16.23}$$

$$C_j(z,\zeta^*;\zeta,\zeta^*) = C_j(\zeta,z^*,\zeta,\zeta^*) = 0;\ j = 1, 2, \ldots$$

Using the analysis of lemma 16.1, it is easy to show that C exists, is analytic for all complex, t, $z,\zeta \in D$, $z^*,\zeta^* \in D^*$, and satisfies

$$\mathcal{L}[C] = 0 \tag{16.24}$$

$$C(z,z^*;\zeta,\zeta^*,0) = 0. \tag{16.25}$$

Hence $V(z,z^*,t) = \mathcal{U}(z,z^*,t) - \mathcal{U}(z,z^*,0) + C(z,z^*,\zeta,\zeta^*,t)$ satisfies $\mathcal{L}[V] = 0$, $V(z,z^*,0) = 0$ and to prove the theorem it suffices to show that V is analytic in $D \times D^*$ for each fixed t. Without loss of generality assume that $u(x,y,t) \in G(\overline{D} \times T)$ and that D has a smooth boundary., Integrating the identity

$$v_t \mathcal{L}[u] - u_t \mathcal{M}[v] = \frac{\partial}{\partial x}[u_{xt}v_t - u_t v_{xt} - a u_t v] + \frac{\partial}{\partial y}[u_{yt}v_t - u_t v_{yt} - b u_t v] + \frac{\partial}{\partial t}[cuv + a u_x v + b u_y v] \tag{16.26}$$

over $D \times T$ gives

$$\iiint_{D \times T} (v_t \mathcal{L}[u] - u_t \mathcal{M}[v])\,dx\,dy\,dt = \iint_{\partial(D \times T)} (u_{xt}v_t - u_t v_{xt} - a u_t v)\,dy\,dt - (u_{yt}v_t - u_t v_{yt} - b u_t v)\,dx\,dt + (cuv + a u_x v + b u_y v)\,dx\,dy. \tag{16.27}$$

Now let $T_0 = \tau$, $v = S$ and $u = V$ and replace $D\times T$ in (16.27) by $D\times T - \Omega\times T$ where Ω is a thin cylinder surrounding the singular line $r = 0$. Note that $V = 0$ on $t = 0$, $S = 0$ on $t = \tau$. Computing the residue as Ω shrinks onto $r = 0$ gives

$$0 = 2\pi \int_0^\tau V_t(\xi,\eta,t)dt + \int_0^t\int_{\partial D}[(V_{xt}S_t - V_tS_{xt} - aV_tS)dy$$

$$- (V_{yt}S_t - V_tS_{yt} - bV_tS)dx]dt$$

or

$$V(\xi,\eta,\tau) = -\frac{1}{2\pi}\int_0^\tau\int_{\partial D}\Big[(V_{xt}S_t - V_tS_{xt} - aV_tS)dy$$

$$- (V_{yt}S_t - V_tS_{yt} - bV_tS)dx\Big]dt, \tag{16.28}$$

and the theorem now follows from lemmas 16.1 and 16.2

Example 16.1: In Theorem 16.1 it is not possible to remove the assumption that $\mathcal{U}(z,z^*,0)$ is analytic. Consider the special case of equation (16.1) when $\underset{\sim}{M} = \underset{\sim}{L}$ and $\gamma = 1$. Then $\underset{\sim}{u}(x,y,t) = e^{-t}\underset{\sim}{u}(x,y,0)$ is a solution of (16.1) and is not analytic unless $u(x,y,0)$ is.

We now turn our attention to constructing a reflection principle for solutions of equation (16.2).

Integrate the identity

$$W_t\mathcal{L}[\mathcal{U}] - \mathcal{U}_t\mathcal{M}[W] = \frac{\partial}{\partial w}(\mathcal{U}_{tz^*}W_t - \alpha\mathcal{U}_tW)$$

$$= \frac{\partial}{\partial z^*}(\mathcal{U}_tW_{tz} + \beta\mathcal{U}_tW) + \frac{\partial}{\partial t}(\alpha\mathcal{U}_zW + B\mathcal{U}_{z^*}W + \gamma\mathcal{U}W) \tag{16.29}$$

over a three dimensional cell $G \subseteq D\times D^*\times T$, set $W = A$, and let $\bar{\mathcal{L}}[\mathcal{U}] = 0$. By lemma 16.1 and Theorem 16.1 the derivatives in (16.29) are well defined. We arrive at

$$0 = \iint_{\partial G}(\mathcal{U}_tA_t)_{z^*}dz^*dt - \iint_{\partial G}\mathcal{U}_t(A_{tz^*} + \alpha A)dz^*dt$$

$$+ \iint_{\partial G}\mathcal{U}_t(A_{tz} + \beta A)dzdt + \iint_{\partial G}(\alpha\mathcal{U}_zA + \beta\mathcal{U}_{z^*}A + \gamma\mathcal{U}A)dzdz^*. \tag{16.30}$$

Now (paying attention to the boundary conditions (16.8) and (16.9) satisfied by A) first let G be the parallelpiped with base $(\zeta_0,\zeta_0^*,0)$, $(\zeta,\zeta_0^*,0)$, $(\zeta,\zeta^*,0)$ and $(\zeta_0,\zeta^*,0)$ and heigh τ, and then let G be the wedge with base $(\zeta,\bar{\zeta},0)$, $(\bar{\zeta}^*,\zeta^*,0)$, $(\zeta,\zeta^*,0)$ and height τ. This yields two equations analogous to equations (11.6) and (11.7) of section 11, and following the analysis of section (11) we arrive at the following analogue of Lewy's reflection principle for elliptic equatio

<u>Theorem 16.2 ([14])</u>: Let $D\times T$ be a simply connected cylindri domain in the half space $y < 0$ whose boundary contains a portion σ of the plane $y = 0$. Let $u(x,y,t)\in G(D\times T)\cap C^2(\bar{D}\times T)$ b a solution of $\mathcal{L}[u] = 0$ in $D\times T$, and on σ suppose $u(x,0,t)= \rho($ where $\rho(x,t)\in \underline{C}'(D\cup\sigma\cup D^*\times T)$ and for each fixed $t\in T$ is an analy function of z and z^* in $D\cup\sigma\cup D^* \times D\cup\sigma\cup D^*$. Then $u(x,y,t)$ can be uniquely continued as a solution of $\mathcal{L}[u] = 0$ in class $G(D\cup\sigma\cup D^*\times T)$ into all of $D\cup\sigma\cup D^*\times T$.

<u>Remark</u>: For more recent results on the analytic theory of pseudoparabolic equations see the University of Glasgow Ph.D. thesis of W. Rundell and the Indiana University Ph.D. thesis of S. Bhatnagar.

References

1. S. Bergman, Integral Operators in the Theory of Linear Partial Differential Equations, Springer-Verlag, Berlin, 1961.
2. S. Bergman and M. Schiffer, Kernel Functions and Differe tial Equations in Mathematical Physics, Academic Press, New York, 1953.
3. L. Bers, F. John and M. Schechter, Partial Differential Equations, Interscience, New York, 1964.
4. R. P. Boas, Entire Functions, Academic Press, New York, 1954.
5. B.D. Coleman, R.J. Duffin and V.J. Mizel, Instability, Uniqueness and nonexistence theorems for the equation $u_t = u_{xx} - x_{xtx}$ on a strip, Arch. Rat. Mech. Anal. 19 (1965), 100-116.

. D. Colton, Cauchy's problem for almost linear elliptic equations in two independent variables, J. Approx. Theory, 3 (1970), 66-71.

. D. Colton, Improperly posed initial value problems for self-adjoint hyperbolic and elliptic equations, SIAM J. Math. Anal. 4 (1973), 47-51.

. D. Colton, Integral operators for elliptic equations in three independent variables, I, Applicable Analysis 4 (1974), 77-95.

. D. Colton, Integral operators for elliptic equations in three independent variables, II, Applicable Analysis, to appear.

0. D. Colton, Bergman operators, for elliptic equations in four independent variables, SIAM J. Math. Anal. 3 (1972), 401-412.

1. D. Colton, On the analytic theory of a class of singular partial differential equations, Analytic Methods in Mathematical Physics, R.P. Gilbert and R. Newton, editors, Gordon and Breach, New York, 1970, 415-424.

2. D. Colton, On the inverse scattering problem for axially symmetric solutions of the Helmholtz equation, Quart. J. Math. 22, (1971), 125-130.

3. D. Colton, Pseudoparabolic equations in one space variable, J. Diff. Eqns. 12 (1972), 559-565.

4. D. Colton, On the analytic theory of pseudoparabolic equations, Quart. J. Math. 23 (1972), 179-192.

5. A. Erdelyi, Singularities of generalized axially symmetric potentials, Comm. Pure Appl. Math. 9(1956), 403-414.

6. A. Erdelyi, W. Magnus, F. Oberhettinger and F. Tricomi, Higher Transcendental Functions, Vol. II, McGraw Hill, New York, 1953.

7. F. G. Friedlander, On an improperly posed characteristic initial value problem, J. Math. Mech. 16 (1967), 907-915.

8. K.O. Friedrichs, On the differentiability of the solutions of elliptic differential equations, Comm. Pure Appl. Math. 6 (1953), 299-326.

9. B.A. Fuks, Special Chapters in the Theory of Analytic Functions of Several Complex Variables, American Mathematical Society, Providence, 1965.

20. P.R. Garabedian, Partial Differential Equations, John Wiley, New York, 1964.

21. R.P. Gilbert, Function Theoretic Methods in Partial Differential Equations, Academic Press, New York, 1969.

22. R.P. Gilbert, On the singularities of generalized axially symmetric potentials, Arch. Rat. Mech. Anal. 6 (1960), 171-176.

23. R.P. Gilbert and C.Y. Lo, On the approximation of solutions of elliptic partial differential equations in two and three dimensions, SIAM J. Math. Anal. 2 (1971), 17-30.

24. P. Henrici, A survey of I.N. Vekua's theory of elliptic partial differential equations with analytic coefficients, Z. Angew. Math. Physics, 8 (1957), 169-203.

25. P. Henrici, On the domain of regularity of generalized axially symmetric potentials, Proc. Amer. Math. Soc. 8 (1957), 29-31.

26. C.D. Hill, Parabolic equations in one space variable and the non-characteristic Cauchy problem, Comm. Pure Appl. Math. 20 (1967), 619-633.

27. C.D. Hill, A method for the construction of reflection laws for a parabolic equation, Trans. Amer. Math. Soc. 133 (1968), 357-372.

28. L. Hormander, Linear Partial Differential Operators, Springer Verlag, Berlin, 1964.

29. S. Karp, A convergent 'farfield' expansion for two dimensional radiation functions, Comm. Pure Appl. Math. 14 (1961), 427-434.

30. P.D. Lax, A stability theory of abstract differential equations and its application to the study of local behaviour of solutions of elliptic equations, Comm. Pure. Appl. Math. 9 (1956), 747-766.

31. H. Lewy, On the reflection laws of second order differential equations in two independent variables, Bull. Amer. Math. Soc. 65 (1959), 37-58.

32. B. Malgrange, Existence et approximation des solutions de equations aux derivees partielles et des equations de convolutions, Ann. Inst. Fourier, 6 (1956), 271-355.

33. C. Müller, Radiation patterns and radiation fields, J. Rat. Mech. Anal. 4 (1955), 235-246.

34. L.E. Payne, On some non well-posed problems for partial differential equations, Numerical Solutions of Nonlinear Differential Equations, Donald Greenspan, editor, John Wiley, New York, 1966, 239-263.

35. C. Pucci, Alcune limitazioni per le soluzioni di equazioni paraboliche, Ann. Mat. Pura Appl. 48 (1959), 161-172.

36. R.E. Showalter and T.W. Ting, Pseudoparabolic partial differential equations, SIAM J. Math. Anal. 1 (1970), 1-26.

37. E.C. Titchmarsh, Theory of Functions, Oxford University Press, London, 1939.

38. B.L. Tjong, Operators generating solutions of certain partial differential equations and their properties, Ph.D. thesis, University of Kentucky, Lexington, 1968.

39. I.N. Vekua, New Methods for Solving Elliptic Equations, John Wiley, New York, 1967.

40. C. Wilcox, A generalization of theorems of Rellich and Atkinson, Proc. Amer. Math. Soc., 7 (1956), 271-276.